Christoph Winter

Contractor-Led Procurement

GABLER EDITION WISSENSCHAFT

Baubetriebswirtschaftslehre und Infrastrukturmanagement

Herausgegeben von Professor Dr.-Ing. Dipl.-Kfm. Dieter Jacob
Technische Universität Bergakademie Freiberg

Für internationales Zusammenwachsen und Wohlstand spielt gutes Infrastrukturmanagement eine zentrale Rolle. Erkenntnisse der baubetriebswirtschaftlichen Forschung können hierzu wichtige Beiträge leisten, die diese Schriftenreihe einem breiteren Publikum zugänglich machen will.

Christoph Winter

Contractor-Led Procurement

An Investigation of Circumstances and Consequences

With a Foreword by Prof. Dr.-Ing. Dipl.-Kfm. Dieter Jacob

Deutscher Universitäts-Verlag

Bibliografische Information Der Deutschen Bibliothek
Die Deutsche Bibliothek verzeichnet diese Publikation in der Deutschen
Nationalbibliografie; detaillierte bibliografische Daten sind im Internet über
<http://dnb.ddb.de> abrufbar.

Dissertation Technische Universität Bergakademie Freiberg, 2002

1. Auflage Oktober 2003

Alle Rechte vorbehalten
© Deutscher Universitäts-Verlag/GWV Fachverlage GmbH, Wiesbaden 2003

Lektorat: Brigitte Siegel / Annegret Eckert

Der Deutsche Universitäts-Verlag ist ein Unternehmen der
Fachverlagsgruppe BertelsmannSpringer.
www.duv.de

Umschlaggestaltung: Regine Zimmer, Dipl.-Designerin, Frankfurt/Main

Gedruckt auf säurefreiem und chlorfrei gebleichtem Papier

ISBN-13: 978-3-8244-7947-4   e-ISBN-13: 978-3-322-81614-6
DOI: 10.1007/978-3-322-81614-6

Foreword

This is the second title in a series which is aimed at giving access as widely as possible to the results of our research into construction business management. We wish to express our gratitude to Gabler to have put their faith in us.

Christoph Winter, who was the first research assistant to join the newly created chair, combines knowledge and practical experience of Anglo-Saxon construction management with awareness of the German construction environment. The Anglo-Saxon countries have not experienced a period of reconstruction, as has occurred after the reunification of Germany, and have thus experienced life-threatening competition much earlier. Contractors in offering producer-led and collaborative procurement have developed instruments to strengthen their competitive position. These are the thoughts that this work now introduces into Germany.

As a true compendium this work is not only directed upstream at the client, but also downstream, adopting a resource orientated look at integrating subcontractors and other suppliers most effectively into the supply chain, in order to facilitate contractor-led procurement in the first place, despite the difficulties of seasonal one-off and prototype production at ever changing locations. Midstream relationships in form of collaborative working or partnering are being considered as well.

Christoph Winter has come to the newly founded chair with this idea already in mind and has continuously developed the theme during all that time. In one move he has highlighted the entire topic of producer-led and collaborative procurement as a means for competitive advantage in the construction industry. It is a truly pioneering piece of management science for the German construction industry, especially in respect of marketing issues and supply chain management. He has done so in a knowledgeable and professional manner with practical applications in mind. All in all it is an unusually robust and substantial piece of work of a senior professional and is not merely a product from the heights of academia. He has succeeded in presenting this most important topic for the construction industry from a national and international perspective in a concise manner. We can only wish for the work to find as many readers as possible. The publication of this dissertation as a book by Gabler Edition Wissenschaft is a significant step in the right direction.

Prof. Dr. Dieter Jacob

Preface

The construction industry is facing an increasingly competitive environment the world over, where greater pressure for change on the existing procedures of the construction industry is brought about by the more powerful of its clients. Clients, who themselves are confronted by the effects of globalisation.

The urgent need for organisations of the construction industry, whether they are contractors or consultants to adapt to their environment in terms of organisational structures and strategy sets the backdrop against which the appropriateness of contractor-led procurement, its circumstances and consequences is to be presented.

This book considers the circumstances that bear directly and indirectly on a contractor's competitive position in variety of construction markets in terms of the roles that various clients take in demanding construction services, in terms of consultants' influence on the construction development process and in terms of the role that the supply chain must fulfil for the accomplishment of successful construction projects.

It appears that meeting clients' demands for a ready purchase of design, procurement and management of construction from a single source is most successfully managed by the adoption of a producer–led procurement path, especially when expecting high levels of efficiency, cost certainty, productivity and quality levels. At the same time a competent main contractor will need to know when it is safe to single source from a supplier, when it is appropriate to undertake joint ventures or when preferred or market place supplier tendering is the most effective method of sourcing a construction project. He has to optimise business relationships and early involvement with his supply chain and be expert in handling consultants, specialist contractors and material suppliers as not only befits a single, but a succession of projects for a variety of clients and project types.

To this end tools are presented that allow a suitable selection of procurement route to be made from a client's perspective and determine the preferred business relationships between main contractor and his supply chain. These are neither too prescriptive nor complex as to prevent their use in every day practical situations and will be of help in an increasingly dynamic and complex construction market.

I would like to thank my colleagues and students at the Chair of Construction Business Management of the Freiberg University of Mining and Technology for their help and valuable feedback when writing the dissertation, and also many other members of the construction industry who in discussion have contributed in some way. I am especially indebted to my supervisor Prof. Dr. Dieter Jacob whose encouragement, assistance and support has made this work at all possible. My thanks are also due to Prof. Dr. Margit Enke and Prof. Dr. Bernd Kochendörfer for their support and critical, yet positive, reviews. Finally, I wish to express my gratitude for the patience and tolerance of my wife, Helga, and two sons, Paul and Joseph, for their support and encouragement during the preparation and writing of this work.

Christoph Winter

Contents

List of Figures and Tables XIII
List of Acronyms and Abbreviations XV

1. Introduction.. 1
  1.1 Abstract.. 1
  1.2 Objective 2
  1.3 Purpose... 3

2. The Construction Industry and its Participants... 5
  2.1 The construction environment, market, industry and its participants.......... 5
    2.1.1 The environment, market and industry in general....................................... 5
    2.1.2 Construction environment 10
    2.1.3 Construction markets.. 12
    2.1.4 Construction industry 18
    2.1.5 Participants of the construction industry 19
  2.2 Development trends in the construction industry structure........................ 26
    2.2.1 Clients... 26
    2.2.2 Consultants 36
    2.2.3 Contractors 41
  2.3 Analysis of trends in the construction industry structure........................... 52
    2.3.1 The pressures of change 52
    2.3.2 Analysis of clients' behaviour and the consequences 53
    2.3.3 Analysis of contractors' behaviour and the consequences 56

3. Overview of Construction Procurement Types 65
  3.1 Introduction.. 65
  3.2 Differentiating between procurement systems and generic procurement techniques 67
  3.3 Types of procurement systems.. 68
    3.3.1 Classification of procurement types.. 68
    3.3.2 Presentation of procurement types 70
    3.3.3 Design-led tendering... 72
    3.3.4 Management-led tendering 77
    3.3.5 Producer-led tendering.. 84
  3.4 A guide to the procurement selection process 88
    3.4.1 Problems encountered during selection 88
    3.4.2 Organisational features of projects 89
    3.4.3 Management approaches for determining selection criteria 90
    3.4.4 Client criteria.. 92
    3.4.5 Project criteria.. 94
  3.5 A general procurement selection model.. 95

3.6 **Standard construction contracts** 99
   3.6.1 Standard documents in the United States 100
   3.6.2 Standard documents in the United Kingdom............................... 102
   3.6.3 International standard documents... 103

4. Contractor-led Scenarios.. 106

4.1 **Introduction** 106
4.2 **Organisational features of contractor-led procurement**........................... 107
4.3 **Positive features of contractor-led procurement**..................................... 110
   4.3.1 Positive features of contractor-led procurement
         in respect of time.. 110
   4.3.2 Positive features of contractor-led procurement
         in respect of cost 113
   4.3.3 Positive features of contractor-led procurement
         in respect of quality... 117
4.4 **Less favourable circumstances of contractor-led procurement**................ 121
   4.4.1 Circumstances less favourable for contractor-led procurement
         in respect of time.. 121
   4.4.2 Circumstances less favourable for contractor-led procurement
         in respect of cost 122
   4.4.3 Circumstances less favourable for contractor-led procurement
         in respect of quality... 125
4.5 **Appropriate application of contractor-led procurement**........................... 128
   4.5.1 Analysis of positive and less favourable features of
         contractor-led procurement... 129
   4.5.2 Two examples of Design and Build projects................................ 133
   4.5.3 Preferred application of contractor-led procurement 137
   4.5.4 Some references to German contracting practice....................... 144

5. The Relationships of a Design and Build Contractor with other

Participants 147

5.1 **The Design and Build contractor and clients**... 147
   5.1.1 General comments concerning the contractor–client relationship 147
   5.1.2 Experienced clients and the concept of "partnering" 154
   5.1.3 Inexperienced and occasional clients 158
   5.1.4 Public sector clients... 160
5.2 **The Design and Build contractor and consultants** 161
   5.2.1 The relationship between contractor, consultants and designers 161
   5.2.2 Good practices for the relationship between contractor and consultants. 163
   5.2.3 Alternative approaches for design completion............................ 165
5.3 **The nature of contractor to contractor relationships** 171
   5.3.1 Types of contractor relationships 171
   5.3.2 Specialist contractors and subcontractors.................................. 175
   5.3.3 Current nature of main contractor-subcontractor relationships................ 178

6. Working with Subcontractors 188

  6.1 Issues to consider when working with subcontractors 188
    6.1.1 The need for subcontracting 188
    6.1.2 Risk management in procurement 189
    6.1.3 The ideal and limitations of early supplier involvement 192
  6.2 Selecting the right governance structure for main contractor-supplier
    business relationships 199
    6.2.1 Current approaches 199
    6.2.2 Procurement classification and strategies for a Design and Build
        contractor 200
    6.2.3 A Design and Build contractor's procurement choices 204
    6.2.4 Systematic approaches to procurement market research 206
  6.3 Behaviour and control exercised in contractor-led procurement 207
    6.3.1 Good tendering and estimating practice 207
    6.3.2 Competitive versus negotiated supplier selection 213
    6.3.3 Supplier appraisal and development 213
    6.3.4 Early involvement tools 214
    6.3.5 Control of the project development process 217

7. Review 220
  7.1 Summary 220
  7.2 Conclusion 223
  7.3 Outlook 225

References 227
Appendix 239

List of Figures and Tables

Figure 1: Elements of industry structure...6
Figure 2: National Diamond 7
Figure 3: Influence of design changes on costs 28
Figure 4: Trends in construction procurement spend 1995-2005 30
Figure 5: Services offered by BDP 39
Figure 6: Proportion of all contractors according to size class...................47
Figure 7: Growth in number of contractors according to size class 48
Figure 8: Development of turnover per employee according to size class.................49
Figure 9: Procurement types 70
Figure 10: Traditional procurement structure...73
Figure 11: Management-led tendering / Construction Management 79
Figure 12: Management-led tendering / Management Contracting.................80
Figure 13: Design and Build 85
Figure 14: General procurement selection model...97
Figure 15: Framework for the selection of consultants by a Design and Build
           contractor..170
Figure 16: Prioritisation and management of risks.......................................190
Figure 17: Interrelationship between risk, cost, design and contract management 191
Figure 18: Procurement classification 201
Figure 19: Example of procurement strategy choices...................................205
Figure 20: Preferred marketing approach to procurement...........................207
Figure 21: Relationship between bid frequency and bid cost......................212

Table 1: Clients' drivers for construction projects in UK 21
Table 2: Market share and firm size indices (in the UK) 46
Table 3: Ownership composition of subcontracting firms (in the USA)..................49
Table 4: Terminology for different procurement groups.............................69
Table 5: Procurement paths suggested for further analysis.........................98
Table 6: Comparison of work by different procurement methods 107
Table 7: Correlation between procurement types.......................................145
Table 8: Advantages and disadvantages of design completion methods 169
Table 9: Correlation between subcontractor selection procedures
          and type of main contract 182
Table 10: Comparison between results and the recommendations of the code.........184
Table 11: Ranking of statements received by subcontractors 185
Table 12: Reasons for subcontractors' difficulties......................................186
Table 13: Factors that influence subcontractors' willingness to bid 187
Table 14: Procurement choices 202
Table 15: Main contractor's subcontractors and suppliers........................205
Table 16: Number of tenders recommended 209
Table 17: Tendering times..210

List of Acronyms and Abbreviations

%	per cent
approx.	approximately
bn.	billion
BOT	Build, Operate, Transfer
e.g.	for example
etc.	and the rest
GDP	Gross Domestic Product
Ibid.	in the same place
IT	Information Technology
m.	million
p.	page
pp.	pages
QS	Quantity Surveyor
UK	United Kingdom
US	United States
USA	United States of America

1.   Introduction

1.1   Abstract

Major clients of the construction industry have been found to organise construction work into fewer, but larger, contracts with more transfer of risk and responsibilities in response to a change from a sellers' market to a buyers' market, and facing a greater choice of procurement methods than ever before.

Main contractors and consultants alike are moving towards multidisciplinary teams offering design and management services, challenging single service consultants or contractors and are in competition with each other over who is leading the process. A consolidation of firms at the upper end of the industry can be witnessed in order to access a wider market and new clients, and at the lower end a specialisation into specific skills or locations takes place, while medium sized firms are increasingly struggling to survive.

A general procurement model serves to identify the appropriate procurement approach for construction needs, as neither clients or construction service suppliers represent a homogenous market. Clients' demands for a ready purchase of design, procurement and management of construction from a single source have been found to be met most appropriately by contractor-led procurement under most, but not all, circumstances, particularly in respect of higher levels of efficiency, cost certainty and punctuality among other benefits.

The consequences faced by a contractor in the leading role of the procurement process are significant, especially in terms of integrating and co-ordinating the entire supply chain to the satisfaction of the client and for anticipated repeat business. This is the chief factor of competitive strength for the struggle of long term survival. A classification model of procurement strategies in respect to parameters of supply risk, strategic importance and frequency of spend offers a tool for the appropriate choice of business relationship with different suppliers.

It is to be anticipated that the future will see an intensification of the changes in the processes of construction procurement described and analysed, which may vary in extent from one market to another, but not in direction.

1.2   Objective

The construction industry is facing an increasingly competitive environment the world over, where greater pressure for change on the industry's procedures is brought about by the more powerful of its clients, who themselves are confronted by the effects of globalisation. The urgent need for firms of the construction industry, whether they are construction organisations or consultants, to adapt to their environment in terms of organisational structure and strategy sets the backdrop against which the appropriateness of contractor-led procurement, its circumstances and consequences, is to be investigated.

The nature of the circumstances that bear directly and indirectly on a contractor's competitive position in a variety of construction markets has to be considered in terms of the roles that various clients take in demanding construction services, in terms of consultants' influence on the construction development process and in terms of the role that the supply chain must fulfil for the accomplishment of successful construction. The effects that these changes have on the participants of construction, who are clients, consultants, contractors and suppliers, must be identified and the consequences analysed.

Within this context, the contractor as producer and in the leadership position of the construction procurement process, from first contact with the client to the completion and possible operation of a building or facility, must master the effective organisation and handling of the supply chain including design, which must be addressed in terms of the benefits and difficulties that face client and contractor alike.

1.3 Purpose

The need to arrive at a meaningful and substantiated outcome to the question of whether contractor-led procurement in construction is appropriate under the actual circumstances that participants of the industry are confronted with and their possible consequences, requires that a large body of literature from a variety of sources and origins is referred to, aimed to achieve an up-to-date account of the situation in the construction industry in some of the major construction markets of the world.

The environment, the market and the industry, together with the participants of construction are described, and placed in context to create a position from which judgement can be passed on the trends in the construction industry and the consequent changes that occur as a result in the behaviour of its key players: the clients, consultants, contractors and suppliers.

The array of procurement types in construction is explained, divided into three distinct groups and presented together with a general procurement selection model based on client and project criteria, which is illustrated by a number of worked examples. The importance that a variety of standard contracts has on the procurement process in the United States, the United Kingdom and internationally is referred to.

Concentrating on the benefits and less favourable aspects of contractor-led procurement in respect of time, cost and quality, the circumstances for the successful application of this type of procurement path are analysed, preferences stated and supported by two brief examples of actual projects. Parallels to German contracting conventions are drawn.

The consequences as a result of such an approach to construction procurement on the relationship between contractor and other participants, especially its range of suppliers including subcontractors, is investigated and, since the ability of organising and co-ordinating the supply chain is of prime importance, a number of issues, not least the selection of the preferred types of contractor-supplier business relationships, are examined.

The overall effect, designed to achieve an objective description and evaluation that compares the application of contractor-led procurement in contrast to other procurement methods, is supported by an examination of the benefits and difficulties associated with such an approach. As a result the appropriate application of contractor-led procurement and the consequences thereof are examined in the light of what is required to have a construction client fulfil his construction needs as efficiently as possible.

2.   The Construction Industry and its Participants

2.1   The construction environment, market, industry and its participants

2.1.1   The environment, market and industry in general

All organisations are faced by an environment that includes everything considered to be outside of a company that either affects it directly or indirectly. While those aspects that define the environment can be described as simple and static or complex and dynamic, in practice the situation encountered is one that ranges from low to high levels of complexity and dynamism[1].

A distinction is made between an industry, an arbitrary boundary within which firms are in competition within each other producing products or providing services, and a market that is any organisation where buyers and sellers are in close contact to determine the price of a product. "Industry" is a supply side concept, while "market" is a demand side concept. There are related industries that produce products and services that share customers, techniques or channels, but they have their own unique requirements for competitive advantage. In practice, drawing industry boundaries is essentially a matter of degree[2]. A framework of competitive forces that determine an industry structure and largely explains the behaviour of its members is influenced and shaped by an interdependent relationship with its market, as shown in the illustration on page 6.

The strength of each of the five competitive forces is a function of industry structure, which is relatively stable, but can change over time as an industry evolves influenced both by environmental forces and firms' strategies[3]. Industry structure determines who captures the value created by firms for buyers, where, for example, the *threat of entry* determines the likelihood that new firms will enter an industry and compete away the value, either passing it on to buyers in the form of lower prices or dissipating it by raising the cost of competing. The *power of buyers* determines the extent to which they retain most of the value created for themselves. The *threat of substitutes* determines the extent to which some other product or service can meet the same buyer needs and thus

---

[1] see, for example, Walker, 1996, pp. 56-78.
[2] Porter, 1990, pp. 33.
[3] Porter, 1985, p. 7.

places a ceiling on the amount a buyer is willing to pay for an industry's product. The *power of suppliers* determines the extent to which value created for buyers will be appropriated by suppliers rather than by firms in an industry and finally, the *intensity of rivalry* acts similarly to the threat of entry[4].

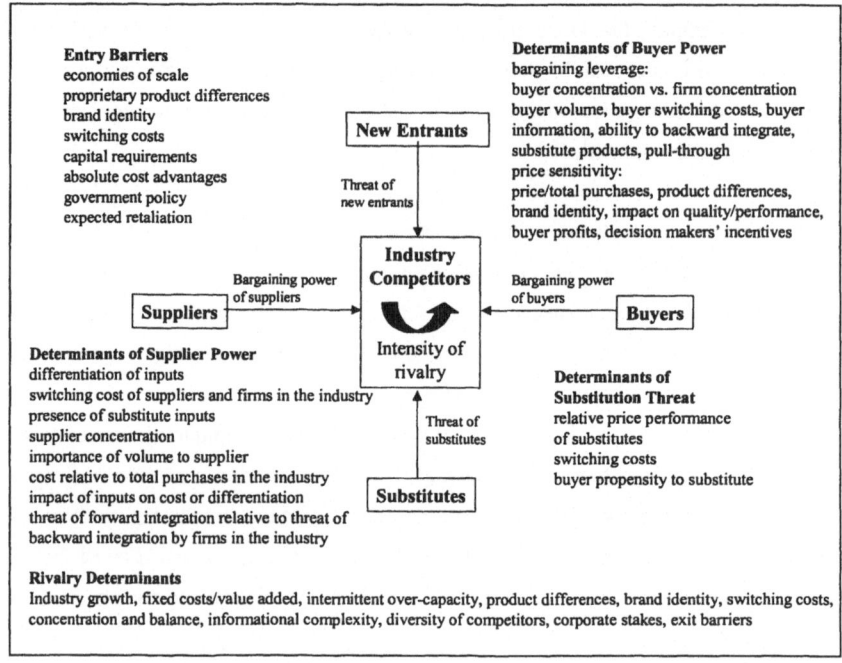

Figure 1: Elements of industry structure[5]

The nature of a market, particularly on a national scale, which is to explain the characteristics and possible success or failure of an industry, has been carefully analysed and described by Porter[6], who has derived the "national diamond" model referring to a system of four broad attributes that shape the environment and market in which an industry competes and is further influenced by two additional variables of government and chance events, as represented diagrammatically over the page.

---

[4] Ibid. pp. 8.
[5] Ibid. p. 6.
[6] Porter, 1990, pp. 27.

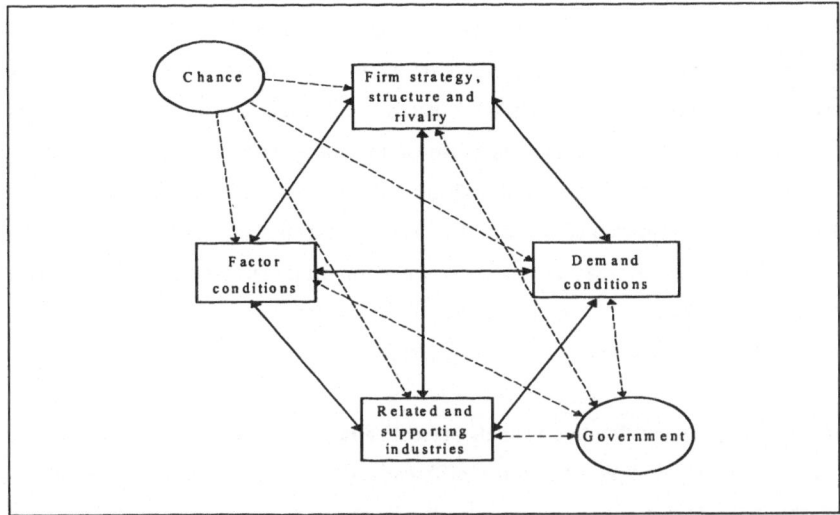

Figure 2: National Diamond[7]

The "diamond" is a mutually reinforcing system, with the effect of any one determinant being contingent on the state of the others and a market which displays a favourable combination of these in respect of a particular industry is likely to advance it across national boundaries. The four determinants can be briefly described as follows:

- Factor conditions, which describe a nation's position in factors of production, such as skilled labour or infrastructure (physical or services), necessary to compete in a given industry.
- Demand conditions, that explain the nature of home demand for the industry's product or service.
- Related and supporting industries, where presence or absence in the market of supplier industries and related industries that are internationally competitive.
- Firm strategy, structure and rivalry, which describes the conditions in the nature of markets governing how companies are created, organised and managed and the nature of domestic rivalry. The latter is described in detail by the elements of industry structure.

---

[7] Ibid. p. 127.

By market structure is normally meant the degree of concentration or distribution of market shares, which is the proportion of all transactions in a product or service involving a buyer and seller. Market structure is also concerned with the extent of product differentiation. By providing a distinct product or service, a firm is in a position to protect its own market share from changes in the prices of other firms, at least to some degree. It is attempted to acquire a niche in the market, a gap not covered by existing products or services provided by competing firms. The purpose being to create a situation in which direct comparisons between products and services is difficult and as a result to be able to increase profit / gross margin.

Differentiation in construction markets presents a rather different problem to the more usual text book example of product differentiation by producers. Under traditional construction contracting systems it is clients with their designers acting as agent and not the producer firms who specify and control the form and content of the build product. This has shifted the focus of attention from the product to the type of service undertaken by contracting firms. In turn, this has led to the description of the product of the construction firm not as the building but as a construction service. As such, it is possible to point out differences between firms competing for the same project in terms of differences in the service provided, although the final building might be identical in terms of its appearance, regardless of the contracting firm employed. However, the rules of simple selective competitive tendering on price alone (as is the case with the public sector in most countries) contains the assumption that all tenderers are equivalent or undifferentiated in terms of the quality of the services they are offering. Differentiation is at best confined to a simple one dimensional sorting of firms into approved and non-approved or tender listed and non-listed firms[8].

Another method that can be adopted alternatively by oligopolistic[9] firms is collusion, where price and output of each producer and the allocation of work and market shares is

---

[8] Gruneberg and Ive, 2000, pp. 91.
[9] Oligopoly refers to a market in which a few firms, which dominate the market, are obliged to enter into a variety of non-price competition such as advertising, promotion and corporate imaging. They create a situation in which direct comparison between products is difficult, by introducing imperfections into the market and product differentiation, in order to increase gross margins. Price competition between oligopolistic firms would be destructive; see: Ibid. p.93.

predetermined by the firms, enabling higher prices to be charged than would otherwise have been the case if normal competition had prevailed.

Barriers to entry are another feature of markets, where firms in a market will tend to set higher margins then they would if they feared the arrival of new entrants. High barriers to entry bring stability to sets of rival firms. Market barriers to entry include[10]:

- Economies of scale, where barriers are associated with the concept of minimal efficient scale of production, below which it would be uneconomic to set up in competition with existing firms and the concept of customer loyalty to existing suppliers or providers.
- Supply chains, where vertical integration or long term contractual arrangements are established with other firms in the production supply chain, which can form effective barriers to new firms.
- Incumbents' cost advantages, where established firms are protected from competition by new entrants if the latter will be unable, on entry, to match the former's level of costs, for example, where experience and learning – curve effects are strong. On the other hand, if there is technological improvement in production methods or in service delivery, existing firms may be left with obsolete, higher cost plant or less effective forms of service delivery.
- Private information, where existing firms possess either private or proprietary knowledge and are in a position to take advantage of information not known by other firms, or is protected by ownership rights, copyright and patents. Established firms may have knowledge about customers, subcontractors, suppliers and competitors that new entrants will not have. This information is not shared and lack of it forms a barrier to entry. Often, this type of technological or market information not generally available outside a firm will actually be known by certain individuals currently employed by that firm. It is open to a new entrant to try and acquire this private knowledge by poaching these employees and represents a problem particularly for firms following strategies of "relational contracting".

---

[10] Ibid. pp.97.

- Client imposed barriers to entry, referring to clients only short-listing tenderers who can demonstrate past experience on similar projects or have been able to pass client's pre-selection criteria.

2.1.2 Construction environment

The process of providing a project is a response to the actions of the environment, which acts in two ways upon the process, indirectly upon the activities of the client of an individual project and directly upon the process itself. At its root, it is the action of forces of the environment on the client's organisation that triggers the need for construction work by responding in order to survive, or to take an opportunity to expand, become more efficient and as a result requiring construction work to be undertaken and providing the construction industry with work. At a strategic level, it will determine how the building should be provided, dependent upon the property market, the technology of the process and may trigger changes to the proposed building required by the client during design or construction. The environmental forces acting directly on the design and construction process can affect the ability of the process to achieve what the client wants. International projects especially have extremely complex environments, where not only the environment generated by the country in which the project is built, but also the environment of the countries providing the construction team and supplies have to be taken into account[11].

The environmental influences acting directly upon the client's organisation, therefore, should determine the organisation structure and mode of operation appropriate to the client's activities. Environmental influences will present opportunities to the client as well and will determine the manner in which such opportunities need to be taken. For example, a client's environment may determine that an additional manufacturing capacity needs a building quickly in order to take advantage of an opportunity. It is thus in the best interest of the client to set up an organisation that is capable of acting quickly to achieve this. If, at the same time, forces indicate a degree of uncertainty of the nature of the market for the product, then the organisation set up to take advantage of the situation must also be capable of achieving the flexibility required. This need may occur

---

[11] Walker, 1996, pp. 62.

at the time of a rise in activity in the building industry, thus creating uncompetitive conditions in terms of price and completion time for the project. The construction process is, therefore, made complex by the type of environment in which it exists, it must produce a clearly defined solution at the technical level of design and construction but must also remain flexible and adaptive to satisfy environmental requirements.

The construction industry's professional and industrial firms, meanwhile, have for many years not adapted a great amount to their environment, as illustrated by the large proportion of projects undertaken predominantly in the conventional pattern in spite of much criticism of this process. The conventional pattern tends to be self-regulating and to function to maintain the given structure of the system, it existing in an environment from which it protects itself. This is achieved through codes of conduct and fee scales of its professional institutions, sometimes established in law[12], which eliminates to a large extent competition between firms, enabling the system to resist change and maintain the status quo. However, the increasingly complex and competitive environment and increasing speed of change in which the system and its clients have to exist have been significant in breaking down such practices, at different rates in different parts of the world, with the Anglo-American sphere of influence leading this process of adaptation. The increasingly multinational nature of the industry's clients and overseas practice have been a major force for change as clients are experiencing novel methods of managing construction procurement to those found at their home base.

At project level, it appears that some clients are adapting by changing the nature of the building process, for example, by introducing the contractor into the design team and so moving nearer to an open adaptive system of construction procurement.

Adaptation at project level takes place in many parts of the world, form the United States across many European States to the Far East and Australasia, especially where large scale projects are underway in response to clients' demands, in respect of traditional institutional domination of the professions and industry firms and their respective representative institutions. This was only possible for projects with clients who themselves were adaptive and not protected in some way from their own

---

[12] For example, the Honorarordnung für Architekten und Ingenieure (HOAI) in Germany.

environments, as opposed to public sector clients, which tend to be protected to some extent.

The process of providing a construction project should be an open adaptive system, but in practice it is always constrained by the environment within it exists, which varies from one market to another. Nevertheless, the process needs to change its structure, if environmental events, acting either directly upon the process or indirectly through the client's organisation, dictate that this should happen[13].

2.1.3 Construction markets

The process involved in determining prices in the construction market is turned upside down, as the client initiates the product and the contractor traditionally has little control over the contract, where price is largely agreed before the contract starts. In addition, the price provides the basic criterion for much of contractor selection. Therefore, the market structure for contracting involves reverse price determination, a reverse auction, with one buyer and competing sellers for a pre-demanded project. This determines the competitive arena, which is client generated and largely of a short term focus for the majority of transactions (i.e. for the duration of a project). Typically, other features encountered in construction markets are a large number of small value orders, extensive division of labour and specialisation of skills, minimal vertical integration and limited advantages of scale, all of which are compatible with industrial fragmentation. Existing barriers are perceived to be low, since[14]:

- low cost of entry is thought to prevail as contractors rely upon human capital.

- expertise and know-how can readily be brought in, the level of which depending on the financial resources available.

- there are presumed to exist only constrained opportunities for limit pricing strategies.

- it is thought that pre-demand limits the effectiveness of marketing, and

- economies of scale are thought unpredictable as characteristics change with each project.

---

[13] Ibid. p. 72.
[14] Gruneberg and Ive, 2000, pp. 97.

The perception of ease of entry to the construction market is in fact only partially correct, and applies only to the lower end of a market of a hierarchical industry based upon subtle forms of entry and exit barriers, which is structured into contractors of varying sizes and a series of project based vertical markets. At the lower end of the market contractors are highly fragmented as described, but as project size and complexity increase and geographical perspective widens, they are more concentrated and management experience and access to financial markets becomes increasingly more important. Only a small number of firms are able to compete at the upper end of the market. Neither is the entry or exit of firms a very common occurrence at the upper end, as contractors have no obvious alternative lines of business to follow, whilst potential new entrants seem discouraged by the need to incur sunk costs for building up expertise and reputation in order to enter the market and cannot be easily recouped at exit. They are probably also discouraged to enter by the large number of existing competitors and unfavourable long term demand trends[15]. What is common, however, is for a construction firm already established in one market or set of market segments to try to enter new market segments when margins there seem to be more attractive and to withdraw from certain segments when margins are relatively low. The example of large UK contractors withdrawing from conventional competitive tendering and targeting negotiated or PFI[16] style projects instead serves to illustrate the point[17]. This tends to equalise margins between construction market segments in the long run.

Construction is undoubtedly in most parts of the world a saturated market nowadays[18], with the exception of some specialist services and expertise such as proprietary process technology, management expertise in delivering large scale and complex projects or offering BOT[19] style services. A saturated market creates intense pressure to push down prices, introduce new features, improve product or service performance and provide other incentives for buyers to replace an established approach with newer, modified versions. Saturation escalates local rivalry, forcing cost cutting and a shake out of the

---

[15] Ibid. pp. 163.
[16] Private Finance Initiative, which generally speaking refers to BOT style projects undertaken in the UK.
[17] Building, 7/9/2001, p. 23; Walter, 1998.
[18] Kommission der Europäischen Gemeinschaft, 1997, p. 7.
[19] Build, Operate, Transfer, for further explanation refer to section 3.3.5.

weakest firms. It also propels domestic firms to look at international markets, as did European construction firms at the end of the post World War II reconstruction phase[20].

As well as domestically, there is no single interrelated construction market at the international level, but a highly stratified and segmented range of market segments. Features that differentiate the construction market at the international level include:

- Size of demand in a particular market, usually by project type, which governs market entry of competitors.

- Access to finance as a key competitive weapon in international construction.

- Explicit and implicit barriers to entry, including cultural, pre-qualification procedures and licensing or bonding needs for contractors.

- Political stability influencing markets and determining the degree of risk in undertaking a project.

- Competitive analysis, which may be difficult for reasons of unknown ownership structures of competitors and suppliers, which is an important issue for a foreign firm.

It is not markets in less developed countries that are important to contractors as a rule, other than as supplier, of unskilled or semiskilled labour, but other construction markets of advanced nations are of significance on account of their market size and investment volume. The nature of the home market as influenced by its resources is a major determinant of competitive advantage for an industry and its members in competition abroad and subsequently at home[21].

It is not only the upper end of the construction market defined by large scale and complex projects, but also regional, local and specialist fragmentation into specialist contractors[22] that limits the number of firms in any one market. Therefore, in some markets relatively few construction firms dominate and control a significant proportion of market sales. Firms are usually well aware of who its close rivals are in a market

---

[20] Porter, 1990, p. 96.
[21] also refer to: Porter, 1990.
[22] For further information, refer to: 2.2.3 and 5.3.2.

segment and monitor their behaviour and success carefully. Much the same is true of markets for professional services/consultants in construction. In both cases, geographical barriers to entry are sufficiently high to exclude non-local firms from the local market for all but the largest projects[23].

In times of strong construction demand contractors' prices are far from stable. Firms tend to raise their prices even before their costs begin to increase, in order to take advantage of a sellers market, but also in anticipation of increasing input prices during the cause of a contract. Similarly, in demand slumps, contractors' prices can fall more rapidly than the index of construction costs[24]. Profit mark-ups can fall to nought percent and in severe competition for work firms have been known to accept negative mark-ups in order to "buy" work to give some cash-flow and turnover, for as long as variable costs are covered and at least a contribution is made to fixed costs. This usually puts added pressure on subcontractors and labour to reduce their charges and wages. This is partly explained by the fact that they do not buy their inputs in fix price markets (e.g. subcontractors' services), that they do not really carry expensive spare capacity and in the way that construction clients often can take advantage of recessionary conditions by deliberately stimulating tender price competition in a way that ordinary consumers could not. Contract prices vary strongly in line with changes in demand, since the structure of construction firms' costs is dominated more by prime costs than overhead costs[25], thus outweighing any tendency for overhead costs and mark-ups varying with demand and output volume. The main forces working to restore normal mark-ups in periods of either increasing or decreasing demand are the entry or exit of firms respectively in the longer run. However, as already pointed out, neither entry or exit of firms in the market for larger projects taken as a whole is a common occurrence, while for smaller firms, especially for very small firms, the ease of entry erodes any chances of restoring mark-ups to higher levels.

---

[23] Gruneberg and Ive, 2000, p. 158.
[24] Index is made up of building suppliers' list prices and wage costs for construction labour.
[25] Ibid. pp. 158.

The construction market, of course, exists not only of construction firms but also includes the building materials industry, which, unlike contracting, is an example of a national oligopoly[26], where for basic building materials like cement, brick and blocks, roofing tiles, glass, plaster board and ready-mix concrete each national economy normally contains just a handful of major, often international, producers with limited market penetration by imports (since largely controlled by the same global companies, e.g. Lafarge, Heidelberger Zement, St. Gobain). Well publicised examples of explicit price fixing agreement between oligopolies in construction materials have included ready mix concrete price rings or cartels[27]. Generally, markets for building materials and components are fix price markets, where price is largely not altered in response to changes in demand, where the quantities produced and offered for sale will tend to be adjusted, unless manufacturing output under full capacity. A special case exists for a few large, powerful buyers of building materials who conduct confidential price negotiations around the size of bulk purchasing discounts off the published list prices and the length of credit period before payment. Building suppliers are known to be among the most important providers of credit to the construction industry, with the larger more influential firms benefiting most[28].

Another market related to the contracting industry is the property market, which also has a fix price character in part. For commercial property, such as offices and retail, already let, rates are normally fixed for several years at a time and usually to a contractually agreed formulae to determine the amount of upward increase in rents at each periodic review date. However, rentals on newly built property are much more flexible as are capitalised market prices for the purchase and sale of property. At a time of excess demand property prices will rise as competing potential owners outbid each other in an effort to acquire the ownership of a building. In recession, however, the property market does not reduce its prices to a level sufficient to attract tenants or buyers for all property offered on the market. During recession many buildings remain empty, in the hope that a tenant can be found within a period of time. A reduction in

---

[26] See page 8 for definition of oligopoly.
[27] Ibid. p. 164.
[28] refer also to sections 6.2.2 and 6.2.3 for procurement characteristics and classification.

potential rental income reduces the value of a property and thus a reduction in rent would, therefore, reduce the book value of assets of a property owning company. The propensity to reduce depends on the financial status of the property owners with those highly geared in cash flow terms[29] having to consider short-term cash flow implications, i.e. secure new rent flows or sales must be obtained, on whatever terms, or else the firm faces bankruptcy.

Over the course of a business cycle[30] profits in any part of the economy will in part be determined by the behaviour of fixed and flexible prices[31]. During recovery phases flexprices will tend to rise more rapidly than fixprices, but conversely, during recessionary phases flexprices will tend to fall faster relative to fixprices. Traditional main contractors and subcontractors are a prime example of construction firms who sell in a flexprice market but buy a large part of their inputs, materials and components in a fixprice market. Thus, these firms' profit margins move strongly pro-cyclically[32]. Management contractors are quite a different case, because they do not buy as many, if any, material inputs and leave that function to their supply and build work package subcontractors[33]. Thus, they buy and sell in flexprice markets and are in a better position to manage their profit margins throughout the difficult phases of the business cycle, whereas traditional contractors with a large directly employed work force suffer badly in recessionary periods. Their potential strength in recovery phases, however, is not easy to protect, with poaching of key staff and personnel well known in the construction industry. The more flexible and responsive labour and supplier markets become, the easier it is to witness this effect.

---

[29] Debt charges per annum high as a proportion of total rental income.
[30] Economic or business cycles consist of alternate peaks of expansion and contraction of demand and output, known as the recessionary and recovering phases.
[31] Characteristics of flexprices (flexible prices) and fixprice (fixed prices) markets are combined in real economies, where markets for commodities and services tend to be flexprice and markets for manufactured goods to be fixprice. However, this simple relationship between fixprice and flexprice does not cover all cases. Where one-off, unique projects are concerned, negotiations between a single supplier and a single buyer will often determine the price on the basis of the personal negotiating ability and bargaining strength of the participants, see: Ibid. pp.50.
[32] also: Kommission der Europäischen Gemeinschaft, 1997.
[33] Gruneberg and Ive, 2000, pp. 166.

2.1.4   Construction industry

While in the past there may have been marked differences in the structure of
construction industries when viewed in terms of national markets, they have become
remarkably similar among the majority of advanced nations. The elements of industry
structure according to Porter happen to be rather similar in most respects in most
advanced nations, probably since key supplier industries, such as construction plant
manufacturers and building material producers, have essentially become multinational
companies offering their specialist equipment, products and services to all buyers
world-wide. Domestic markets are no longer capable in supporting a sufficient number
of key customers that are large or capable enough to matter. A similar environment in
terms of entry barriers, determinants of supplier power, rivalry, substitution threat and
buyer, is encountered in nearly equal measure in all advanced nations. Differences that
do occur are usually not sufficiently significant to substantially change the overall
situation, although national business cycles do not necessarily occur concurrently the
world over[34].

National statistics are difficult to compare accurately with one another, thus the
following information is to act as a guide only, but is to serve as a means to indicate the
overall structure and breakdown of construction industries across some world markets.
For example, new construction output in the United States accounted for approximate
8% of GDP, with around another 5 % for refurbishment work[35]. In the United Kingdom
the figure for total construction output is about 8 % of GDP and for Germany a figure of
approximately 11 % for construction output is quoted[36]. The average European Union
total construction output of its 15 member states combined amounted to 11 % of GDP,
where construction in this instance includes all residential, commercial and engineering
work as well as all supply networks and all activities at every stage of construction from
concept stage via feasibility studies, design, design detailing to construction, including

---

[34] e.g. the boom in construction after Germany's reunification at a time of world recession during the
early 1990's.
[35] Levey, 1999, pp. 1; Halpin and Woodhead, 1998, pp. 13.
[36] Building, 11/01/2002, pp. 36-47.

its maintenance followed by demolition and finally recycling or disposal of waste materials[37].

In nearly all countries the construction industry is fragmented, where the distribution of size of construction firms follows a highly skewed pattern. The smallest construction firms in size occur with the largest frequency. As the number of employees in each size range increases, the number of firms per size range declines. However, in terms of total turnover or employment accounted for by all the firms in each size range something quite different appears, with an increasing share of either as firm size ranges rise. In the United States 86 % of all firms in the construction sector employed less than 5 people, but the remaining number of firms accounted for 80 % of total output[38]. In the United Kingdom a similar picture emerges, where 84 % of firms employed 3 people or less and only 1 % employing more than 35 people[39]. In Germany 76 % of all firms employed up to 10 employees and 6 % of firms employed more than 50 people[40]. As can be seen, the skew may vary in severity from one market to another, however, exhibit the same basic shapes.

2.1.5 Participants of the construction industry

Whilst focusing on the contracting organisations of the construction industry they do not operate in a vacuum and rely on the client who demands construction services. At the same time they have to arrange themselves with specialist consultants on the one hand, including architects, a variety of engineers, project managers and quantity surveyors, and on the other hand, specialist trade firms and suppliers of building materials and components. The following presents an overview of the participants and a description of their characteristics.

Clients

The range of building clients is extensive from central, regional and local government, public organisations and housing associations to private organisations in many shapes

---

[37] Kommission der Europäischen Gemeinschaft, 1997, pp. 1.
[38] Levey, 1999, pp. 1; Halpin and Woodhead, 1998, pp. 13.
[39] Seely, 1997, p. 2.
[40] Syben, 2000.

and sizes with a wide variety of buildings including factories, warehouses, educational and health facilities, offices, entertainment, retail and residential, water, energy and communication infrastructure, roads, rail and many other smaller specialist projects. Three basic groups of clients can be identified amongst this variety to aid an understanding of clients and their behaviour in terms of construction needs[41]. These are:

- The individual client, who tends to be the exception nowadays for any but the smallest projects, particularly where he is to be both owner and occupier. Examples are a couple preparing to have a house built for themselves or a sole owner of a business. Even at this relatively simple level the way the construction team obtains the information it needs must depend upon understanding the client's activities, organisation and relationships.

- The corporate client, which includes all companies and firms controlled other than by a sole principal. These are a group of small, simply structured organisations to the massive multinational corporation. The myriad of functions, sizes and structures of firms in this group poses particular problems for the construction team.

- The public client, which includes all the publicly owned organisations that have the authority to raise finance to commission construction work. In all such cases the funds will normally be raised by taxation or in the money market on the authority of the government. Many of the features that apply to the corporate client are applicable to the public client as well, but the situation encountered is often more closely constrained and difficult through having to work through committees whose authority may not be clearly defined, and the need to be seen to be accountable to the public on monies spent.

Clients' policies and procedures vary considerably, but they all perceive the need to procure construction services, this need driven by wanting to increase capacity, upgrade buildings, meet business and strategic objectives and to expand into new markets. The table on the next page is a result of research into client drivers for construction projects in the United Kingdom[42]:

---

[41] Walker, 1996, p. 83.
[42] Gibb and Isack, 2001.

upgrade facilities	17.4 %
reduce operating costs	17.1 %
add capacity	16.9 %
health and safety	12.6 %
expand by geographic region	11.4 %
legislation	10.5 %
expand into new markets	8.0 %
other reasons (including making a profit, expanding property portfolio, relocating facility and facilitating new technology)	0.6 %

Table 1: Clients' drivers for construction projects in UK

The most important feature of any building project should be the client's objective in embarking on the construction of the project, which results as a basic response to environmental forces in order to survive, or above this level, to respond in order to expand as a result of drive and motivation. Survival as the basic objective of clients can be defined as maintaining their position relative to those of their competitors. This is more easily conceived for commercial organisations, but is also true for public clients[43]. The client as employer usually pays the cost of the work, but, traditionally, does not usually come into contact with the various members of the construction team apart from the architect or perhaps a professional project manager, although he is very much concerned with all that it involves.

Consultants

This group of participants of the construction industry are commonly referred to in part as the design team, typically consisting of architect and engineers, including structural and services engineers or any other engineers for specialist fields such as fire precaution, acoustics, lighting, landscaping, etc.

The architect has traditionally been regarded as the leader of the building team, but inroads of both project managers and other professionals are tending to change their

---

[43] Walker, 1996, pp. 89.

role[44]. The architect often receives the commission to design and supervise the erection of a building, but the degree of specialised knowledge required for the design of a modern, complex building is such that he will need the assistance from specialists, who typically include structural, mechanical and electrical (services) engineers. Some parts of the world, especially the United Kingdom and Commonwealth countries, rely on the services of a quantity surveyor to advise on contractual and cost aspects, to prepare bills of quantities and other contract documentation[45]. Architects may also need advice on ground investigation, landscaping and other aspects, such as remediation of contaminated sites. The architect in his classical role acts as an expert advisor and agent for the employer. Here, although he is primarily a designer, he is nevertheless involved in the production of a building from inception to completion – from pre-design through production drawings and details to supervising the contractor. He traditionally assumes the important task of co-ordinating the activities of everyone else involved in the project. It is widely recognised that, with the complexity of modern buildings, construction techniques, employers requirements and vastly increased number of people involved in the preparation and execution of the work, the architect's traditional role is virtually impossible to accomplish on almost any project but the most basic[46].

Construction cost is a key problem area for a client, who has commissioned an important building or engineering project. In respect to costs, it is a quantity surveyor who, as a cost and contract administration expert, has the prime task to ensure that the project is kept within the agreed budget on behalf of the client and that he obtains value for money.

Consulting engineers, specialists in structural design and mechanical and electrical engineering services, prepare the necessary designs, specifications and other relevant documents, obtain quotations for the work and subsequently supervise the work on site under the overall control of the architect in a traditional set-up. The structural engineer must ensure structural efficiency and stability and at the same time he will minimise

---

[44] Seely, 1997, p. 37.
[45] For more information on the role of the Quantity Surveyor, refer, for example, to: Seely, 1997; Winter, 2000, pp. 63-74.
[46] Seely, 1997, p. 39.

considerable obstructions by structural members and assist in producing a logical systematic construction process. Engineering services are concerned with the control of the internal environment by means of heating, ventilating, air conditioning and lighting installations and providing utilities such as electrical supplies, lifts and compressed air. The proportion of capital costs denoted to services varies considerably with building design and function, but typically they account for between one and two thirds of total expected costs. Other consultants who may be engaged include landscape architects, interior designers, acoustic consultants or fire protection experts, depending on the type of project and elements required.

The large number of consultants involved in the preparation and control of the design and construction process, often independent units which are, however, interdependent in terms of the work they have to undertake and the considerable variety in the range and quality of skills they offer, makes for a complex situation. This is further compounded by the variety of clients and projects overlying the professional relationships[47], calls for the management of the contributors to the project[48]. Traditionally, the architect served in this role, or in the case of some large firms and the public sector an "in-house" capability served this function. Nowadays, however, with increasing complexity and dynamism as well as increasing specialisation both of clients and construction organisations concentrating on their own business objectives, it is becoming less likely that objectives of the firm, the project and the individuals will be satisfied simultaneously, thus calling for the need of professional project management. Its sole objective should be that of the satisfactory completion of the project on behalf of the client and normally crosses, therefore, a firm's boundaries and for its purposes temporary management structures are created for the duration of the project[49].

Contractors
Very basically, a contractor is any organisation which has accepted legal responsibility for executing certain specified works in return for payment. Now, there is a great variety

---

[47] see also chapter 3.2.
[48] Walker, 1996, pp. 104.
[49] However, see also comments regarding unbiased, neutral advisor in section 3.4.1.

of organisations fulfilling such a role, ranging from the one person firm to a large multinational company with many thousand employees. Also, the value or extent of the work in any one project, for which responsibility was accepted, varies greatly, with responsibility for all works at one end of the spectrum to only a very small part of all works to be carried out at the other end. A BOT[50] style approach represents the maximum extent of responsibility for a contract including design, construction, operation and maintenance and a sub-contract, for example, for diamond-core drilling of a few holes in a concrete wall represents the minimum extent in respect of a single project.

One commonly refers to a contracting organisation under a building contract which has accepted responsibility for the execution of the whole of the works in return for payment as the "main contractor". The main contractor who is responsible for the whole of the construction works and uses a combination of subcontractors and directly employed labour is known as a general contractor and the method of contracting as a whole is known as "general contracting"[51].

An organisation with responsibility for some part of the construction work, whether with or without a design input, under the employ of a main contractor is referred to as "subcontractor". Often, the term is also used to cover those organisations with a subsidiary relationship such as "works contractors" under the employ of a "management contractor".

"Management contracting" is a particular method of contracting in which the main contractor (now known as the management contractor) carries only a low risk (for cost and time on behalf of the client i.e. "for fee") and where the actual work is carried out by works contractors, also referred to as subcontractors by some, directly contracted with the main contractor.

---

[50] See chapter 3.3.5 for an explanation of Build, Operate, Transfer (BOT).
[51] Hughes, Gray and Murdoch, 1997, p. 10.

The system of contracting by which the work is carried out by specialist trade contractors, who are contracted directly with the client, is referred to as Construction Management and the organisation employed to manage and control the construction of the client's project, in co-operation with the designers and other consultants, through the trade contractors is called the construction manager. Often, the services of management contracting and construction management are performed by main contracting organisations, who have parted from all of their direct employed labour force in recent years in response to an increasingly variable and dynamic economic environment.

Specialist trade contractors are firms, who offer and execute a specialism in any or all of design, manufacture, purchase, assembly , installation, testing and commission of items that go into the construction of a building. Those firms can be separated into "specialist contractors", who offer and execute a design service for the item they manufacture / select / purchase and install for the construction of a building, and "trade contractors", that offer and execute work of a skilled nature for the construction of a building, but without a design input[52].

Consequently, all specialist and trade contractors can be subcontractors or works contractors, depending on their contractual relationship to the project's client.

A special case in respect of subcontractors and their relationship to the general contractor and client is the issue of nomination as practised in the United Kingdom[53]. This is a process whereby construction clients instruct general contractors to employ specific subcontractors, who usually have been selected before the general contractor and have a close relationship with the architect for the purpose of project design input. Payment for such work is not based on the general contractor's rates, but is awarded by prime cost sums in the tender and contract documents. However, the general contractor is responsible for all operations and instructions and there is no contractual relationship between the client and the nominated subcontractor just as is the case with domestic

---

[52] Ibid. p. 10.

[53] In recent years the practice of nomination has declined for a number of reasons, primarily for reasons of complications based on the three-way relationship between client, contractor and subcontractor.

subcontractors, who have been selected by the general contractor directly without any involvement of the client or architect[54].

2.2   Development trends in the construction industry structure

2.2.1   Clients

Clients' perception of the construction industry

An influential client in the UK was quoted to have said that "the construction industry is too complex, costs me too much money and does not deliver what I expect it to deliver"[55]. Generally, it is believed that clients are often confused by an increasing number of participants with each person in the construction team wanting authority over the project, but no one willing to take financial responsibility. At the same time clients question the relevance of traditional professional boundaries (between architects, engineers, construction managers and contractors) and challenge the worth of many functions[56].

The conventional or traditional procurement method, referring to general contracting with design separation) is increasingly considered to be unsuitable, as too long and difficult when related to industrial and commercial buildings, but requires enormous pressure to change since it has become institutionalised within the industry[57]. The same can be said for separate trades contracting in countries where it represents the traditional contracting method, as it does in Germany.

A survey of a cross section of large construction clients in the UK in 1999[58] revealed that clients generally held mixed views of contractors' performance, with 60 % expecting to have projects delivered to agreed time or costs, or both, and about one third still expected either time or costs, or both, to increase. Less than 10 % of clients expected their projects to be delivered early or under budget.

---

[54] For more detailed information about the relationship between main contractor and subcontractor see chapter 6.
[55] Seely, 1997, p. 17.
[56] Ibid. p. 515.
[57] Ibid. p. 517.
[58] Chevin, 1999.

It can be said with some confidence and without much criticism from clients of construction services that the construction industry is held in fairly low esteem the world over, be it the United States, the United Kingdom, France, the Netherlands, Germany, the Far East or Australia. In all of these countries clients, whether public or private, are as a rule not particularly satisfied with the results of the construction industry in terms of either cost, time or quality.

Clients' own behaviour

It can be argued that many of the problems encountered by clients in the construction industry are down to their own behaviour when procuring buildings. For example, often a mismatch between selected procurement methods and client expectations and characteristics occurs, a situation that has been created by institutionalised attitudes and a lack of strategic overview[59]. Another contention is that too many changes are introduced when a scheme is already underway. This stems, it is argued, from an inadequate brief from the client to the consultant and/or contractor, which subsequently requires detailed changes in specification as the client decides what he actually wants. These changes have severe implications for both costs and programme as shown overleaf[60 61].

Clients often place too much emphasis on the lowest price only, which will not necessarily be secured from a competition for a lowest bid price alone. Particularly the public sector, on the basis that public accountability is the reason to award a contract to the lowest bidder with quality seen as a "given" covered by the specification and the contract, have experienced large cost and time overruns with construction projects[62]. In the United Kingdom, the Levene Report, issued by the Cabinet Efficiency Office, based on an investigation of 20 major government projects, revealed that costs had increased by 24 % and many were overtime. Another survey claimed that on 803 government

---

[59] Seely, 1997, p. 517.
[60] Cox and Townsend, 1998, p. 24.
[61] CIOB, 1999, p. 14.
[62] e.g.: Cox and Townsend, 1998, p. 24; Seely, 1997, p. 521; Kubal, Miller and Worth, 2000; Blecken, 1998.

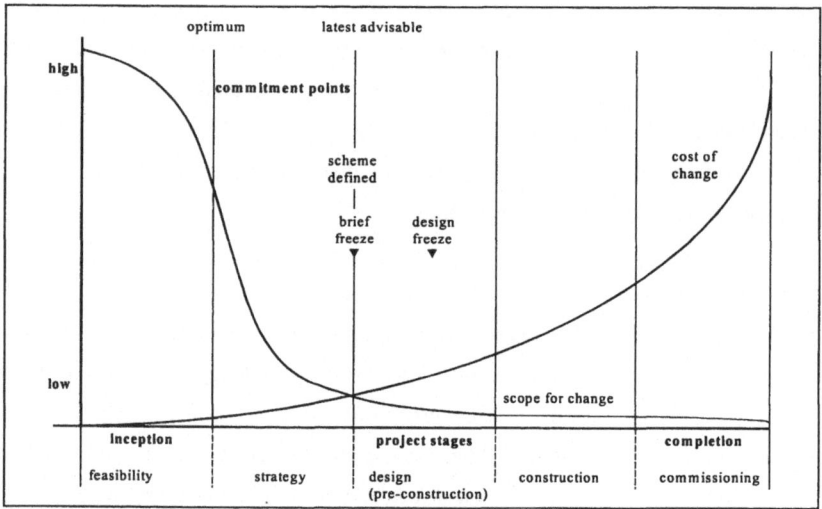

Figure 3: Influence of design changes on costs[63]

construction schemes in 1993 – 1994 more than one quarter finished over budget and nearly one-fifth were late with two-thirds of civil engineering schemes finishing late[64].

In Germany reports of the federal ministry of construction and a number of state audit offices, have indicated that the public administration in charge of construction development, design and construction is much too inefficient and as a consequence exceedingly expensive. Taking all cost factors arising under the public administration of construction projects into account, it has been stated that additional costs in the order of 40 % to 50 % occur, compared to corporate activities[65].

There is a view that too many clients see construction services as a commodity only, although they are in most cases not off-the-shelf products, but are highly specialised and thought intensive services that must be tailored to each individual project. However,

---

[63] Ibid. p. 14.
[64] Seely, 1997, p. 522.
[65] Blecken, 1998.

clients often regard these services as commodities and base their selection on low prices only from a short list of firms that meet minimum qualifications[66].

Clients have increased their use of risk transference to other parties by means of legal contracts, making an onerous allocation of risk in an attempt to reduce their own burden either by imposing risks upon the contractor that are best carried by the client, or by not providing for proper reimbursement of risks carried by the contractor. This is often presumed to be an effective contractual means to resolve exposure to risks during a project's construction by assuring that a client would not have to pay for this risk allocation. This appears to be fuelled by a desire by some clients to get "something for nothing"[67].

The combination of inadequate briefs, emphasis on low bid price only and onerous risk transfer without proper consultation or reimbursement has given rise to numerous problems in cost, time and quality during the course of project completion and has resulted in contractual conflict fuelling an adversarial culture. Summarising the effects that clients' behaviour creates and causes problems, are:

- the lack of a clear contract strategy,
- inadequate briefing and planning,
- improper assessment and inappropriate allocation of risk,
- communication problems throughout the supply chain, beginning with the client,
- insufficient pre-planning,
- inadequate selection and adjudication of tenders,
- traditional forms of contracting creating the potential for conflict,
- a vicious circle of "claimsmanship", and
- payment problems.

Changes in clients' behaviour

A change of behaviour just described can be detected amongst clients in both the United States and the United Kingdom for both private and public sector clients, form short

---

[66] Kubal, Miller and Worth, 2000, p. 17.
[67] Ibid. p. 9; Cox and Townsend, 1998, p. 25.

term, low price and indiscriminate risk transfer to more long term, responsible and appropriate risk sharing.

The shift in clients' behaviour is illustrated by the following results of a survey in the United Kingdom of a cross section of large private and public sector clients in 1999[68]:

- Competitively tendered work accounted for 61 % of the clients' £ 7 bn. workload in 1999, compared with 87 % in 1995. By 2005, it is expected to drop slightly further to 59 %.
- Negotiated or "partnered" work accounted for 16 % of the 1999 workload, a rise of 3% over the past five years and is expected to rise to 18 % by 2005.
- Stakeholder procurement (such as BOT schemes) accounted for 23 % of the 1999 workload, compared with less than 1 % in 1995. It is expected to remain steady at this level.

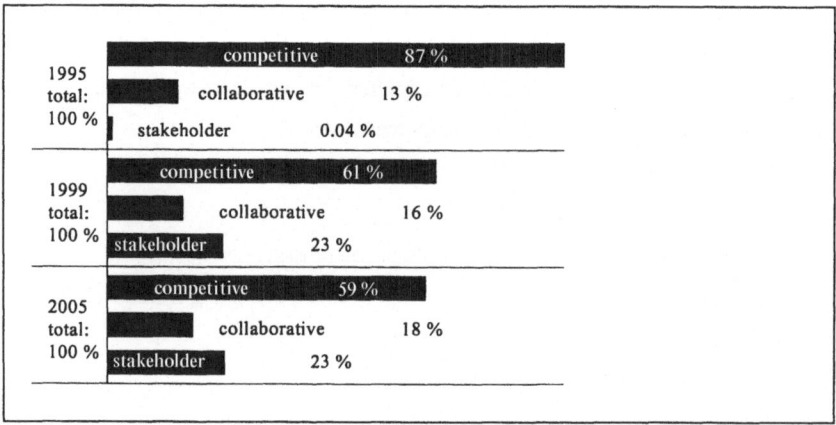

Figure 4: Trends in construction procurement spend 1995-2005

- Nearly two out of three clients expect to use fewer contractors and reduce the number of contractors on tender lists.
- Two out of three clients want to see greater consolidation of top contractors and more foreign firms winning work.

---

[68] Chevin, 1999.

- Nearly half of clients say unjustified contractors' claims have fallen in the past five years and one-third say life cycle costs have fallen.

Clients are discovering that requirements for some type of buildings (e.g. hospitals) are getting so complicated, that the normal practice, whereby the client prepares the brief and appoints the designer, was no longer feasible. It appears that a planning team is necessary to prepare the brief, in order to tackle certain complicated briefing problems[69].

Greater significance is now attached to regeneration and / or renewal of former industrial land and its decontamination and a noticeable shift of emphasis from out of town greenfield development to renewal of town centres is apparent[70].

It is clients that drive the stimulus for innovation in procurement in most cases, as they have in a customer-driven world of falling trade barriers, in which a construction industry with low input costs and high output costs means a competitive disadvantage, slow response to market demand, excessive demands on hard-earned cash and high on-going running costs[71]. With the creation of client groups such as the Construction Clients' Forum (CCF) and the Construction Round table (CRT) in the United Kingdom, the demand side of the industry has recently become less fragmented[72]. Effectively, the gap between what are considered to be small occasional main clients and those that are generally large, regular and experienced clients[73] has substantially grown, with a two tier market of big contractors winning a bigger share of the high value collaborative work and the middle ranking players doing the competitive work where margins are lower[74].

---

[69] Seely, 1997, p. 17.
[70] Ibid. p. 515; Dielschneider, 2000, p. 22.
[71] Cox and Townsend, 1998, preface.
[72] Ibid. p. 18; Tookey, Murray, Hardcastle and Langford, 2001, p. 21.
[73] See also sections 5.1.2 and 5.1.3 for more information on the difference between experienced and inexperienced clients.
[74] Ibid. p. 21; Chevin,1999.

A change from a seller's to a buyer's market and general individualisation has occurred[75] and client groups are now faced with a greater choice than ever before as the industry, like most others, has become global and more complex. The domination of the industry by client groups and generally too many contractors chasing too little work has resulted in demand pressure on price, an upward pressure on quality, an assurance of service and an upward pressure in the ability to meet deadlines, deal with parties and groups involved, undertake client liaison and manage any specific local factors such as traffic disruption, domestic and commercial inconveniences for whatever projects are undertaken[76]. Clients thus demand a shift in attitude from product excellence to service assurance and at the same time they are seeking to organise construction work into fewer and larger contractors, with more risk and responsibility transferred to the contractor[77].

There is a trend for clients to no longer focus on the principle of "lowest price wins" and a move towards a "multi-criteria-selection" approach, indicating that a choice of procurement and contractor selection is made on a value rather than lowest price judgement[78]. Clients have recently truly opened their eyes to the fact that value for money will not necessarily be secured by competition for lowest bid price alone[79]. Evidence suggests, that clients now select contractors by their ability to construct using "preferred modalities"[80] in their approach throughout project delivery, thus seeking a contractor's capability as well as low cost[81]. Most clients do not expect negotiated or partnered work to increase costs[82]. Governments too, have taken a view that procurement of construction projects needs to focus more on quality and value for money instead of lowest price in the short term, as demonstrated by a number of reports revealing overspends and time extensions in traditional procurement. Increasingly, they are basing decisions on whole life costs rather than on initial tenders only. The UK

---

[75] Lahdenperä, 2000, pp. 121.
[76] Pettinger, 1998, p. viii.
[77] Turner, 2000, p. 3.
[78] Wong, Holt and Cooper, 2000, p. 772.
[79] Cox and Townsend, 1998, p. 24.
[80] See section 6.3.4 for examples of early involvement tools in procurement.
[81] Hill, 2000, p. 22.
[82] Chevin, 1999.

government has been advised to change their relationships with industry by working with the best and most co-operative practitioners, but making no compromises with the incompetent or adversarial. The aim is to rectify the lack of accountability, communication failures and over-optimism on budgets that have plagued government's construction projects in the past. Hence, the development of "PFI" procurement methods[83] as well as programmes for improved construction performance such as the Movement for Innovation (M4I) and the Construction Best Practice Programme[84].

It has been claimed that clients in the Netherlands, for example, are not only focusing on financial aspects but also consider integral performance, a life-time approach and integration of services and construction, requiring the combination of design with the construction process[85]. The Government of the Netherlands wants to apply the following principles in the procurement of construction services, including: an output orientated approach, with targets formulated as a set of minimum functional requirements, an enlargement of scope and a choice of optimal procurement approaches, which is to achieve a balance between control and the quality of product[86]

In France, clients demand a combined product and service from the construction sector and look for new forms of co-operation between conception and realisation of new projects. Hence, government too is promoting a process described as "delegated management of public services by private firms under a global contract" based on a form of contract called the "Marché d'Entreprises Travaux Publics (METP)". A government authority awards to a contractor the design and construction of a facility as well as the management of the service which the facility provides for a specified period, in return for regular payments after which it falls back to the authority[87].

Private users and major corporations in the United States have reorganised their real estate holdings into either profit centres, sometimes outsourced, or the holdings are

---

[83] PFI = Private Finance Initiative, see section 3.3.5 for additional information.
[84] Hill, 2000.
[85] Bremer and Kok, 2000, pp. 102.
[86] Ibid. p. 105.
[87] Campagnac, 2000.

implemented and managed at the lowest possible costs. A development that increasingly spreads throughout the economies of the world. Commitments of assistance required by clients not only start earlier but end later as well. The trend indicates a leaning towards performance contracts in construction rather than a completion of a building to drawings and specification only. A substantial increase of Design and Build projects[88] has occurred in the United States, supported by the realisation of clients that they should not have to retain or attempt to transfer design risks but rather that design risk should be retained by the architectural and engineering firms that are paid a fee for their professional services. Design and Build contracting assures the client that all professional risks associated with the design are retained by the construction professional[89]. The federal government has adopted the use of Design and Build procurement for a number of building projects, whereby the selection of the winning consortium is based on competitive review of proposals from each consortia[90].

It is not only general contracting that is used more widely in Germany but also Design and Build and Turnkey contracting that are becoming more widespread in their application, where even the classical public sector is on occasion pursuing construction projects on the basis of output specifications only. It has been estimated that 30 % of project value in Germany is carried out under some form of general contracting, with some of it as Design and Build or Turnkey, as compared to separate trades contracting[91].

Clients' demands from the construction industry

Clients are generally not only organising construction work into fewer and larger contracts, with more risk and responsibility transferred onto contractors, but also resist increases in prices, who are encouraged by studies that have claimed that construction is relatively inefficient and capable of substantial savings[92]. It can be said that clients generally want higher quality buildings at lower prices and which are produced more quickly, coupled with a better service form the construction industry[93]. There is wider

---

[88] For a description of the Design and Build process see section 3.3.5, for project examples see 4.5.2.
[89] Kubal, Miller and Worth, 2000, pp. 9.
[90] Halpin and Woodhead, 1998, p. 73.
[91] Kapellmann, 1997, p. 2.
[92] Turner, 2000, p. 3.
[93] Seely, 1997, p. 519.

recognition that increased quality is achievable and therefore expected on the part of clients, who specifically expect improvements in quality of finished projects and after sales commitment[94]. Increasingly, pressure is put on construction firms to produce higher standards of building, meet the needs of clients and reduce costs[95]. Specifically, clients the world over demand[96]:

- Greater familiarity with clients' businesses and corporate culture.
- Full involvement with establishing and implementing contracting and project strategies.
- Definition of project risks, their size, allocation and management.
- Knowledge about value for money in context of certainty of outcome relative to the inter-linking elements of time, quality and cost.
- Flexibility in responding to clients' requirements.
- Response to clients' needs for responsibility and accountability.

This must be reflected in the ability of construction service providers to control costs within established budgets, to provide strategic construction project financial advice, to understand alternative finance (e.g. BOT), to be aware of environmental issues together with a knowledge of environmental implications of the brief, specification, use of materials and decontamination of land and to display an increased capability and breadth of service generally[97].

Especially, the ready purchase of design, procurement and management of construction from a single source, is felt to be pre-requisite for meeting all these clients' demands[98], who, however, at the same time do not compromise in demanding greater flexibility of the design and construction process while expecting higher levels of quality, efficiency and punctuality[99].

---

[94] Pettinger, 1998, p. 220.
[95] Davey, Lowe and Duff, 2001, p. 1; Lahdenperä, 2000, p. 121; Kapellmann, 1997, p. 4; Turner, 2000, p.3.
[96] Seely, 1997, p. 516.
[97] Ibid. p. 515.
[98] Ibid. p. 515.
[99] Lahdenperä, 2000, pp. 121; Madine, 2001 b).

2.2.2   Consultants

As a result of increasing complexity of the environment in which construction took place during industrialisation, specialisation of the contributors to construction projects has increased throughout the world since the 1800's and early 1900's from the basis of the architect/builder into architects, specialist engineers, quantity surveyors and experts besides contractors in all shapes and sizes. Even within specialist occupations there are often further specialist subdivisions, for instance, the design architect, detailing architect and job architect. In the quantity surveying field there are building economists, bill preparers and final account specialists[100]. There has been a great proliferation of consultants in recent years, with both technological and managerial expertise, reflecting the increasing complexity and dynamism of the construction environment[101].

Differentiation is at its highest on a project when professional consultants are from separate firms and they will be differentiated from the contractor to varying degrees depending upon when and how they were appointed. If positive attempts are not made to integrate them, then the effect upon the project outcome can be serious[102].

The process of adapting to the increasing complexity and dynamism of the construction environment had slowed down as the professions protected themselves from their environment and attempted to maintain the status quo[103]. The process of designing a project on behalf of a client needs to respond to its environment, but during most of the 20th century it has, to a degree, protected itself from its environment by the establishment of codes, procedures and conventions, which have been granted validity by public authorities, professional institutions and other bodies associated with construction[104].

The perception by contractors and clients alike is that architects and other consultants often lack appreciation of the practical implications of their designs and expert advice.

---

[100] Tookey, Murray, Hardcastle and Langford, 2001, p. 75, p. 109.
[101] Pettinger, 1998, p. 27.
[102] Walker, 1996, p. 115.
[103] Tookey, Murray, Hardcastle and Langford, 2001, p. 25.
[104] Ibid. p. 28.

Some construction faults stem from poor detailing and problems can result from the use of new materials, inappropriate usage of materials and from poorly understood sophisticated service components. They have been accused of supplying inadequate details, working to unrealistic programmes and making excessive changes to design during construction[105]. Other criticisms of consultancy practice from within the industry and elsewhere, are[106]:

- Being too prescriptive from their own distinctive way of doing things, but without sufficient attention to clients' specific requirements.
- Sometimes putting great pressure on clients to accept their recommendations, effectively railroading a client into a particular direction without realisation of the consequences.

It is the architect within the field of consultancy who traditionally was, and in many cases still is, regarded as the leader of the building team, although the inroads of project managers and of other professionals are tending to change this traditional approach[107]. It is the complexity of modern buildings, constructional techniques and employers' requirements and the vastly increased number of people involved in the execution of work, which have necessitated a changed attitude and role for the architect. Either architects have to acquire different skills in business and management or increase their specialisation[108].

Leading professionals in years past were primarily experts in their respective disciplines with management skills being a secondary attribute, but this is now changing with success frequently depending on the ability to manage as clients are now seeking single point responsibility for overall project delivery from the professional advisers[109]. Not only is design responsibility in construction projects widening with an increasing use being made of specialist contractors and manufacturers carrying out the design of their

---

[105] Seely, 1997, p. 37.
[106] Pettinger, 1998, p. 27.
[107] Seely, 1997, p. 37.
[108] Ibid. p. 39.
[109] Ibid. p. 527.

own work, but also the environmental influences upon the traditional design process, particularly those being transmitted to it through its clients, have resulted in the process being much more responsive[110].

Professional services are now facing an environment characterised by demand pressures on price, upward pressure on quality, reliability and durability and pressures to extend the quality and nature of relationships. Until very recently professional organisations have existed on a combination of reputation and access based on a relatively steady and assured client bases and known, or almost known, values of work. This has changed as larger consultants and practices appear and have gained market access and share at the expense of smaller organisations[111]. This is the most positive approach of differentiation and integration in practice with the creation of multidisciplinary organisations that employ within the one firm all the professional services associated with construction projects. Ideally, specialists working in project dedicated teams within such organisations create the combination which allows the highest level of integration to occur. However, if such organisations continue to organise in functional departments of specialist skills, a great opportunity to integrate will have been lost[112].

There is a distinctive move towards more multidisciplinary teams offering a design and management service, challenging existing single service consultants and firms[113].There have been many examples in recent years of growth through amalgamation of professional firms and the creation of Design and Build companies acquiring an in-house capacity for designing and contracting projects, in order to be better placed for handling more easily the environment in which they now operate[114]. An example of a very large and successful multidisciplinary practice is Building Design Partnership (BDP), where staff embrace all the construction related professions as illustrated over the page:

---

[110] Tookey, Murray, Hardcastle and Langford, 2001, p. 28-29; Hughes, Gray and Murdoch, 1997.
[111] Pettinger, 1998, pp. 233.
[112] Walker, 1996, p. 115.
[113] Seely, 1997, p. 523.
[114] Tookey, Murray, Hardcastle and Langford, 2001, p.25.

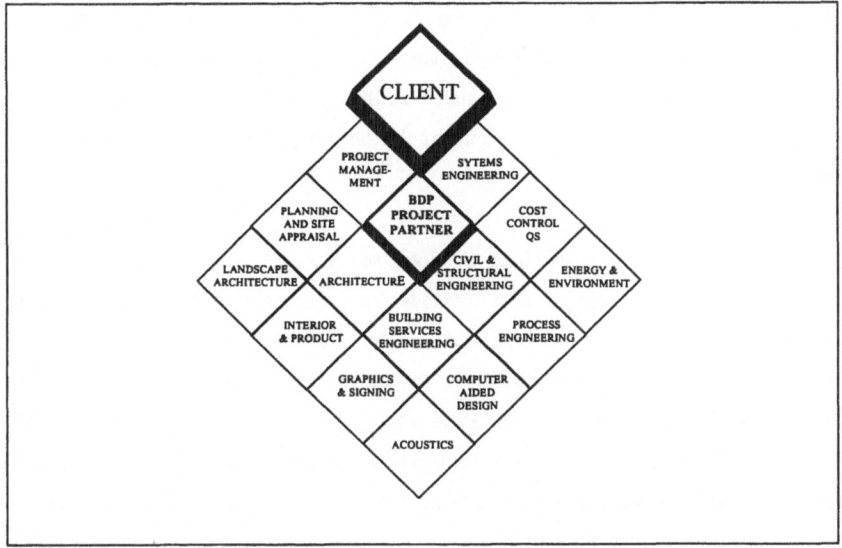

Figure 5: Services offered by BDP[115]

These examples of mergers and acquisitions of consultants with or by a variety of other organisations signal a breaking down of the long established barriers and opens up a new era. initially, practices changed from partnerships to limited companies in order to reflect current business practices generally and made it easier to expand and acquire other businesses. For example, in 1996 the engineering consultancy firm of W.S. Atkins acquired the surveying practice of Faithful and Gould, to create the largest facilities management consultant in the United Kingdom with a combined staff of around 5000, including engineers, architects, quantity surveyors and project managers. Faithful and Gould was the largest firm of quantity surveyors in the United Kingdom with 850 employees, which in 1995 became a limited company and had a 5 year plan to double its size through acquisition with increased support from banks and to increase its profitability by economies of scale. Forty percent of the firm's work was in facilities management with the remainder in quantity surveying and project management. Atkins dominated the public sector having purchased large sections of the government's privatised Property Services Agency (PSA), whereas Faithful and Gould's clients were

---

[115] Seely, 1997, p. 524.

mainly in the private sector, thus complementing each other. Both companies considered their merger to be a positive response to their respective clients' needs and that providing extra services and specialist skills would bring added value to their business[116]. Now, W. S. Atkins are one of the largest players among the UK's PFI market.

As multinational clients are searching for global solutions to their building needs, other examples of firms responding to their demands occur. For example, HOK International, already the biggest architectural firm in the world[117] have forged an alliance with four additional European architects (Altiplan / Brussels, Arte Charpentier / Paris, estudio Lamela / Madrid, Novotny Mahner / Frankfurt) in 2001. The alliance is called HOK Partnerships Network, which enables the firm to work with four more European centres and service an increasingly important type of client, who is the global corporation, increasingly demanding global solutions. The network was developed and built according to the old adage of "think globally, act locally". To illustrate the magnitude of HOK's operations, about 32 % ($ 53 m) of total fee income are from multi-locational work for multinationals, with its top 10 clients generating 69 % of this, e.g. HOK's exclusive global contract with Nortel Networks that has spawned 2000 projects over the past five years with more than $ 5 bn in construction costs and at least $ 100 m in fees.

Other examples include consulting engineer Mott Macdonald, which has 50 offices and 5000 staff around the world and multidisciplinary firm Arup with 70 offices world-wide. Arup operates a global business desk in New York to track current and potential corporate clients, already counting Procter & Gamble, credit Suisse, First Boston, HSBC and Ford among its clients. In order to service major clients consultants such as HOK, Arup and Mott Macdonald have introduced account managers to look after clients, to be always in touch with them and ensure delivery to deadline and quality expectations and have problems discussed and resolved [118].

---

[116]Ibid. p. 527.
[117] Seddon, 2001.
[118] Ibid, p. 46.

Another recent merger has involved Aecom, a US-based multinational company with more than 12,500 staff operating in USA, Europe, Middle east, Asia and Australasia, with consultant engineer Oscar Faber. They will merge with fellow engineer Maunsell, already owned by Aecom, to form a firm with four divisions and 1,800 staff in the UK, thus representing one of the six largest consulting engineers in Britain[119].

It has been stated that most large consultants would like to take their business this way in the long term as it offers the potential for greater consistency of workload in the long run, a refined design and a relationship of trust and respect. However, international expertise has to be matched with local knowledge in order to service demanding, multinational clients, expecting projects to be finished on time to high standards anywhere in the world, which entails handling local cultures, planning regimes, procurement policies and building regulations to produce a product that matches corporate expectations[120].

Another trend can be seen in the increasing move of engineering consultants into the field of management consultants[121]. This type of diversification is to create more stability, allow forms with management consultants to have a wider market appeal, is demanded by clients since it is key clients that want to use the same company for management and engineering as they pursue an approach of single point responsibility. Engineering consultants primarily win management consultants work from their existing client base.

2.2.3 Contractors

A common theme throughout the discussion referring to the development trends of both clients and consultants so far, was the fact that the environment in which the construction industry operates is becoming ever more complex and dynamic and as such is exposed to similar processes as most other industries. Less than a century ago it was customary for most buildings to be designed entirely by one architect and for the

---

[119] Thompson, 2001.
[120] Seddon, 2001.
[121] Madine, 2001 a).

building to be erected by a single contractor (in the Anglo-American world) employing all the necessary craftsmen and labourers, or if operating a separate trades system to be supervised by the very same architect, and assisted by a clerk of works if necessary.

Not only are complexity and dynamism of competition increasing substantially, and the value of time becoming ever more significant, but also modern buildings require greater investment in services, sometimes as much as 50 % of capital investment including information technology facilities and associated sophisticated systems, intelligent low or passive energy buildings becoming more common, and all to satisfy the more searching needs of the occupants, either individual, corporate or public. Together with the use of new materials, components and new techniques and methods of construction as well as increasing mechanisation, there has been a replacement of traditional craft skills in fixing techniques, integration between specialists responsible for structures and claddings to provide a complete shell and work being moved off site to factories. An increasing proportion of building work is in mechanical and electrical services with much greater modularisation of components as one-off design is likely to be replaced by flexible servicing with plug-in components and greater integration of services and dry finishes[122].

Pre-fabrication developments, either modular or volumetric, is particularly suited to hotels, student accommodation, prisons, apartments, warehousing and the education sector, all exhibiting the common denominator that each lends itself to repetition[123]. Off-site production entails construction in a controlled environment protected from the vagaries of the weather and allows much improved quality assurance[124]. At the same time the autonomy of the craftsman is curtailed and there is a concomitant increase in the reliance upon technical documentation due to the increase in technological complexity. Therefore, effective use of new technology relies on the skills of technical designers, where the design team needs to increase their understanding of modern manufacturing technology. If not understood, it cannot be harnessed effectively[125].

---

[122] Seely, 1997, p. 21.
[123] e.g.: Davis Langdon & Everest, 2002.
[124] Ibid.
[125] Hughes, Gray and Murdoch, 1997, pp. 11.

Equally, it is common for as much as half of the construction industry labour force to be engaged in work of building alterations, maintenance and repair in Europe, with apparently around one-third of the labour force involved in such work in newer markets such as the United States or Australia[126].

An environment of greater complexity and dynamism lets firms to specialise further in order to be in a position to manage such complexity, while facing increasing competition. Throughout the construction sector practices and methods are changing, thus calling for new methods of co-operation between all contributors to the construction process with greater flexibility throughout the entire process. Alternative forms of procurement routes and better team work to generate more efficiency, innovation, improved quality and better safety are sought and tried as well as searching for more effective ways of communication and reconfiguring the supply chain of the construction process by integration, improved management and new site techniques[127].

Competitive intensification is transforming the order of priority of contractors' activities. The nature of investment required is, therefore, concerned with creating long-term, continuous profitable contractor-client relationships, transforming services standards and adopting the attitude that the delivery of construction expertise is concerned as much with service as it is with production expertise and building and creating much more comprehensive information and databases in all sectors[128]. Above all, there is a need to reconsider and adapt the rules to which contractors perform in the industry. For example, are they willing to become Design and Build contractors and / or to diversify downstream into maintenance and facilities management? Are they willing to become subcontractors or franchisees? Relatively, the greatest pressure for change may be felt by medium-sized firms, left with fewer opportunities to work as large or small firms[129] as will be described subsequently.

---

[126] Kommission der Europäischen Gemeinschaft, 1997; Levy, 1999, pp. 2.
[127] Kommission der Europäischen Gemeinschaft, 1997, pp. 11.
[128] Pettinger, 1998, pp. 226.
[129] Turner, 2000.

All of these significant factors combined, together with more general technological, political, social and economic changes have accelerated the change from direct employment of a contractor's work force to subcontracting and/or specialist contracting.

The field of subcontractors expanded as design and construction techniques became more sophisticated and more general contractors relied on subcontractors to increasingly perform their work. Even in the United States during the 1950's, when subcontracting became more widespread, some traditional "full service" general contractors, who continued to maintain a large workforce and owned substantial amounts of heavy construction equipment, viewed those contractors who subcontracted all of their work as "brokers" and "business men" but not true builders[130].

Today, it is difficult to find any significant number of contractors who employ teams of skilled workers on their payrolls year round in the United States, as speciality contractors[131] are more skilled, more efficient, more flexible and more competitive in their chosen field of work[132]. The same can be said of the United Kingdom, where it has been stated that it is currently universal practice for the specialised trades skills to be provided by independent trade contractors, or even self-employed individuals and that it is now virtually unknown for a general building contractor to provide building skills from internal, directly employed resources[133].

Although Germany is still the land of separate trades contracting, there is a growing trend of general contracting, estimated at about 30 % of all construction value in the economy[134] and consequently, with an increasing number of general contracting projects, the use of subcontracting spreads. Between 1980 and 1997, there has been a marked increase in subcontracting across firms of all sizes, but especially large firms of over 1000 employees have seen the proportion of subcontracting grow form 25 % to 46% of a firm's total cost structure[135]. In total, subcontracting accounted for 14.4 % of

---

[130] Levey, 1997, p. 2.
[131] In the US specialist and trade contractors are often referred as speciality contractors.
[132] Ibid. p. 2; Kubal, Miller and Worth, 2000, p. 34.
[133] CIOB, 1999, p. 29.
[134] Syben, 2000, p. 125; Kapellmann, 1997, p.2; Sperling, 1999.
[135] Syben, 2000, p. 125; Wischhof, et al., 2000, p.39.

contractors' costs in 1980. By 1997, this proportion had grown to 30.3 % across firms of all sizes. With increasing size of firms, the proportion of subcontracting accounting for a firm's costs rises to 46.1 % in the case of firms of over 1000 employees in 1997. Small firms of between 20-49 employees had a subcontracting rate of only 16.4 % in Germany at that time.

The construction industry of the Netherlands relies heavily on subcontracting and is an important factor even for small firms, where in 1997 the proportion of subcontracting of firms with less than 20 employees accounted for 26 % of a firm's costs, of firms with between 20-100 employees for 39 % and of firms with more than 100 employees this figure increased to 46 %[136].

Subcontracting is thought off in the USA in such terms, that without the expertise and efficiencies displayed by the subcontracting industry construction would undoubtedly be less productive and more expensive. General contractors and the construction industry at large have recognised this fact long ago, as the "full-service" contractor gave way to the "broker"[137], who relies on the subcontracting trades to meet the demands of clients for most competitively priced projects, high quality levels and on-time completion schedules[138]. General contractors have thus shifted from building to management and co-ordination and have sought to extend their role into design and management of the construction process as opposed to the traditional contracting approach[139]. Consequently, building contractors have resolved to become repositories and suppliers of expertise in the management of logistics, human resources, sourcing and finance, other than craft skills[140]. Contractors are thus more likely to be managers and co-ordinators of other companies, which is a tendency taken to its logical conclusion in construction management projects, where a construction manager only co-ordinates and advises. A contractor in the role of a construction manager should ensure that everything necessary for the work is available for each contributor to the

---

[136] Ibid. p. 80.
[137] also: Jacob, 1997; depicts an analogy of general contracting with supermarket trading.
[138] Levey, 1999, p. 5.
[139] Seely, 1997, p. 2.
[140] CIOB, 1999, p. 29.

construction process. This characterisation of general contracting requires for the contractor to take an active part in communication and decisions between the design team and specialist contractors, which places high demands on the skills and technical knowledge of the general contractor in the role of the construction manager in complex projects[141].

Interestingly, there is ample evidence that changing construction markets and restructuring of construction industry causes considerable problems for many firms in the category of medium sized contractors and there are many insolvencies in this group. There is evidence of some degree of polarisation towards large and very small firms[142]. This tendency of medium sized building contractors exhibiting a poorer performance than larger or smaller counterparts can be witnessed in a number of countries including the United States, United Kingdom and Germany, where, for example, between 1981 and 1996 minor and major firms in the UK have increased their outputs per firm by 112 % and 67 % respectively, while intermediate firms have increased theirs' by an average of only 55 %. A trend for consolidation of large firms along with a net loss of market share especially for the middle sector and an increasing number of small firms is apparent.

Variable	Firm size	Year			
		1981	1988-1989	1992-93	1995-96
market share index	small	100	102	110	115
	medium	100	90	88	87
	major	100	112	106	104
relative firm size	small	100	123	178	212
	medium	100	101	140	155
	major	100	128	156	167

Table 2: Market share and firm size indices (in the UK)[143]

---

[141] Hughes, Gray and Murdoch, 1997, p. 79.
[142] Seely, 1997, p. 6.
[143] Stumpf, 2000.

The same trend can be observed in Germany, where medium sized, traditional firms are forecast to fail in the near future and exit the market, fragment into a number of small firms or get taken over. This trend especially affects traditional so called all-purpose contractors with contracting capabilities in a number of sectors, who are in danger of losing out to the process of increasing specialisation and total service delivery from a single source[144]. Some statistics to illustrate this trend[145]:

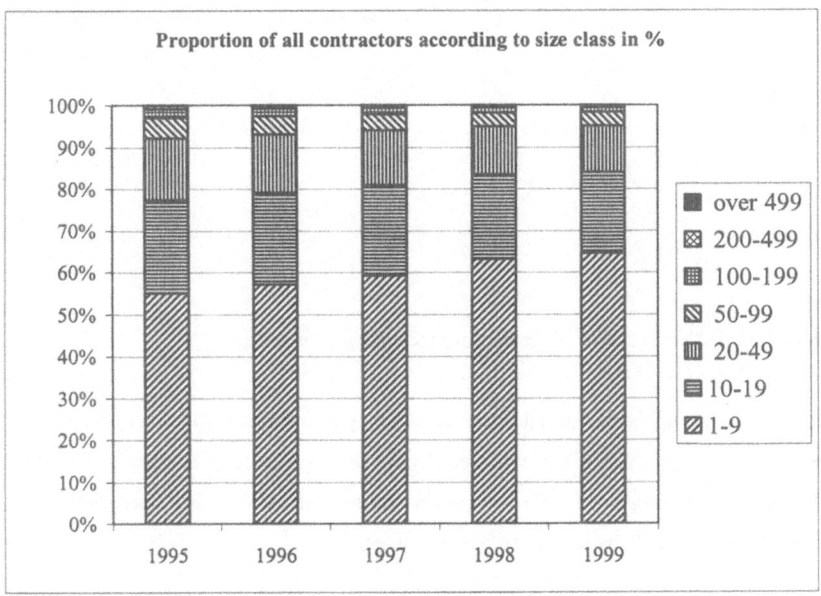

Figure 6: Proportion of all contractors according to size class

---

[144] Wischhof, et al., 2000, p. 119.
[145] Hauptverband der deutschen Bauindustrie.

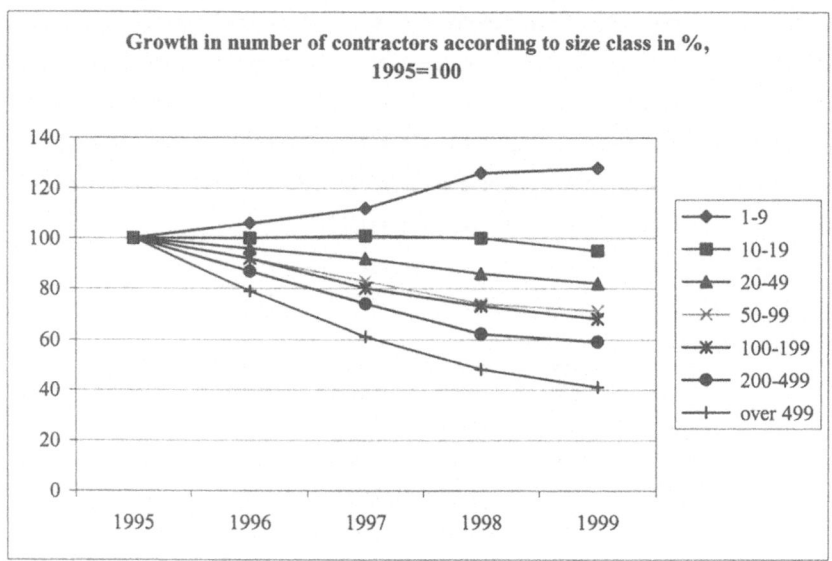

Figure 7: Growth in number of contractors according to size class

Not only is the number of small firms and their proportion of all construction firms increasing in Germany, but turnover per employee of contractors has been increasing and generally the larger the firm the greater the rate of increase has been during the period of 1995 to 1999, as illustrated over the page:

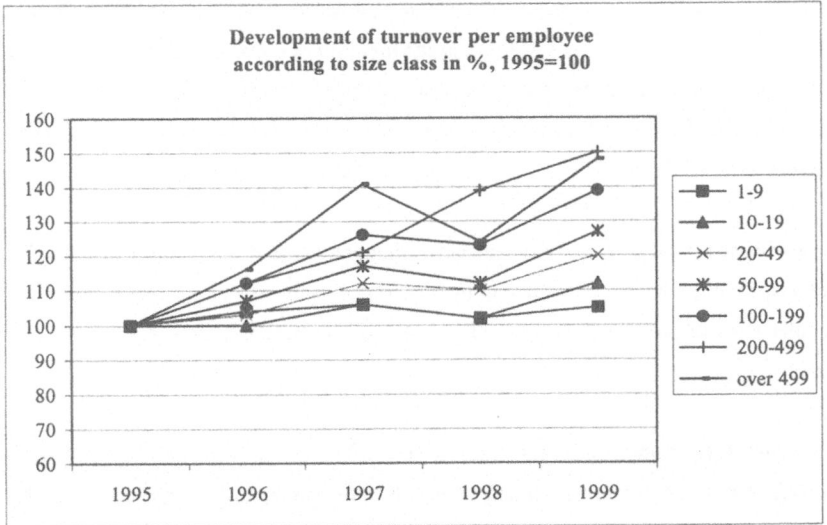

Figure 8: Development of turnover per employee according to size class

These numbers further underline the trend towards a greater concentration of large firms, emphasise the growing number of small firms and that the middle sector gradually looses ground over the long term[146].

Just as a further example of where this trend leads to, some numbers of the US market, where the ownership composition of subcontracting firms has changed over the years from 1992 to 1996 as follows[147]:

Number of owners	1992	1994	1996
1	22 %	22 %	26 %
2-4	54 %	53 %	61 %
5-9	18 %	16 %	9 %
10-29	6 %	8 %	3 %
30 or more	2 %	1 %	1 %

Table 3: Ownership composition of subcontracting firms (in the USA)

---

[146] Refer also to: Schwarz and Schmutzer, 1997.
[147] Levy, 1999, p. 5.

Of course, the number of owners of a firm does not necessarily relate directly to the size of a firm, however, generally speaking, in the majority of cases it can be assumed to be a fair reflection of the actual size of a firm. Therefore, the above figure demonstrates that the proportion of small firms is not only growing, but that the number of very small firms is growing fastest.

Alongside the trend of increasing subcontracting activity and the disappearing middle, there appears to be an acceleration in the consolidation of larger firms, absorbing smaller ones to provide access to either new geographic areas, new markets or new clients[148].

The American market has experienced an influx of foreign construction firms establishing operations in the United States, or, more often, have gained access to this market by acquiring American contractors through outright purchase. These have included in the past Japanese companies, initially ostensibly to service their long term clients, but today have expanded their operations. German construction firms were also attracted by the sizeable market. Holzmann purchsed J.A. Jones and Lockwood Greene Engineering and Hochtief has recently bought the Turner Corporation. Other European firms have included Bovis of Great Britain, who, before having been sold by P&O to Australian's Lend Lease to become Bovis Lend Lease, had taken over Lehr McGovern, and Skanska of Sweden, who took over a number of contractors along the East Coast including Sardonia Construction, W.J. Barney and Beacon Contractors. In 1998, Skanska (USA) ranked ninth among the top 400 contractors in the USA[149].

The international interrelationships of the construction sector is increasing further, whereby specialisation on an international scale will take place with competitive advantage dependent upon domestic factors of the "national diamond". In order to deliver a complete construction service package from a single source (one-stop shopping) international alliances, such as witnessed among consultants, both project based and temporary as well as more substantial and of a long term nature or outright

---

[148] Ibid. p. 4.
[149] Ibid. pp. 4.

take-overs will become more commonplace. This places even more pressure on the medium sized construction firm, who will suffer from increasing specialisation and flexibility of very small firms on the one hand and from management and servicing strengths of large firms on the other hand[150].

For example, large European contractors are increasingly investing directly in other European states, in the USA, South America and Asia, which take different forms of co-operation or merger with local companies and involves the transfer of technology, investment or expertise and represents a preferred method of accessing new markets[151]. Some examples within Europe are HBG's take over of four sizeable UK contractors to form a new group and its purchase of Weiss and Freytag in Germany, itself now a target of the second largest Spanish contractor Dragados, whose aim is to become the third or fourth biggest construction group in Europe[152] with an approximate combined turnover of £ 7 billion[153]; Skanska's take-over of Kvaerner after it had taken over Trafalgar House, the swap of activities between Wimpey and Tarmac, now Carillion, or Walter Group in Germany wholly merging with Heilit & Woerner and Dywidag. Amec, currently the biggest UK contractor according to turnover, has bought the Canadian engineering and technology group Agra as part of a bid to transform itself from a low-margin contractor working on general construction projects into a high-margin, techno-logy based global service partner. The chairman was quoted as having said "Our business goal is to be the preferred supplier and business provider for companies like BP and Shell"[154]. Furthermore, it was claimed that Amec no longer did "hard bid" construction work in the UK and that all work was now "partnered" on a Design and Build or Design, Build and Manage basis[155][156].

---

[150] Wischhof, et al., 2000, p. 108.

[151] Kommission der Europäischen Gemeinschaft, 1997, p. 8.

[152] After Vinci, the world biggest construction firm, operating in more than 100 countries, with an approx. turnover in 1999/2000 of £ 10.6 billion and Bouyges, operating in about 80 countries, with an approx. turnover of £ 8.2 billion in 1999/2000. Hochtief and Skanska follow, each group with approx. £ 7 billion turnover in the same year and similar in size to the new Dragados group. Thompson, 2002.

[153] Building, 8/2/2002.

[154] Seddon, 2001.

[155] Design, Build and Manage equates to BOT .

[156] Building, 7/9/2001.

What has followed the trend to internationalise, is the ability to communicate directly from anywhere to anywhere through electronic connectivity, which has created a global village and for many domestic general contractors has created an opportunity for an entry into new and /or foreign markets[157]. This paradigm is a two way phenomena, which offers opportunity as well as threats to a company's existing service. Domestic firms not already owned by foreign corporations or internationally active, have to emphasise teaming, partnering and forge strategic alliances with these to compete effectively. Not only communication but also the increasing need to compress programmes of a typical building project is forcing the construction industry to implement techniques that reduce the overall time to produce a completed project. Therefore, as the time value of money grows, there is an increasing dependency upon effective computerised scheduling software, ideally tied up with key subcontractors' and suppliers' production control systems. By constantly tracking the components manufacture from design to installation, the contractor can integrate product specific information directly into the overall project schedule to accurately reflect the components delivery and installation data with progress on site[158]. The combination of CAD files, CPM's[159] and all other project files (correspondence, instructions and variations between client, designer, subcontractor, supplier, etc.) into a computer database that is shared by all members of the project team is possible and will soon become standard in the industry[160] [161].

2.3   Analysis of trends in the construction industry structure

2.3.1   The pressures of change

All previous discussion has concentrated on describing the current situation of construction not only from a contractor's perspective but including other very important participants in the construction market, not least the client and consultants. It is thought

---

[157] Kubal, Miller and Worth, 2000, p. 6.

[158] Ibid, p. 35.

[159] Critical Path Method (CPM) refers to linear network scheduling or programming, which describes the sequence of activities that must be performed without delay in order not to compromise the completion date. These activities do not carry any float.

[160] The Building Centre Trust, 2001.

[161] For example Citadon, the largest provider of construction industry portals including project management systems, which was formed by the merger of Bidcom Inc. and Cephren in the United States. McAll, 2000.

necessary to take such an expansive approach to ensure that the relationship that characterises a contractor's behaviour can be analysed in an objective manner. It is particularly helpful in order to make a reasoned argument for the appropriateness of producer / contractor led construction, who must consider his market and supply chain, particularly specialist contractors, who are the key source for any construction activity.

Viewing the elements of industry structure according to Porter[162], it has for many years only experienced a small amount of adaptation to its environment in terms of organisational structure and strategy. Having protected itself from it, it is now facing an increasingly competitive environment in which both the construction industry and its clients have to exist. This has been significant in breaking down barriers not only in isolated parts but across most of the world, as clients have brought to bear greater pressure for change on the industry's procedures as a result of the increased competition which clients themselves are confronted with, often referred to as the effects of globalisation.

2.3.2   Analysis of clients' behaviour and the consequences

To begin with, it is perhaps most important to realise that there is in fact not a single, homogenous group of clients but, as already explained, a vertical hierarchy of markets, which serve the need of a variety of clients. It is important to realise that the trend from a larger contractor's perspective, who is in the market for large scale and complex projects, is towards greater concentration and globalisation, with a handful of clients in each market segment operating on a world-wide scale. The higher the degree of buyer concentration, the less will be the scope for high margins and higher profitability among contractors[163], simply because clients' buying power becomes stronger. This contrasts, for example, with the market for new housing, where there is usually a multiplicity of buyers, each with a negligible market share being served by a large number of regionally based contractors in most countries[164].

---

[162] See section 2.1.1 for methodology.
[163] Gruneberg and Ive, 2000, p. 96.
[164] An exception, for example, is the UK, where on account of a scarce supply of land large, nationwide speculative builders with completion numbers of about 10,000 units a year dominate the housing market.

A further differentiating factor is the degree of importance applying to the construction product within a client's own value chain, which influences the amount of attention paid by the client to the contribution that the construction project is going to make to his own competitive position. With ongoing intensification of competition among large clients of construction services every aspect that influences their own competitive strength gains in significance including construction. It does not matter whether it is an inexperienced client, who purchases a one-off and thus important building product, or an experienced client, who purchases a series of building products, each one not particularly significant but overall just as important to his business strategy, but the demand placed upon the construction process in delivering a completed product in terms of time, cost and quality is increasing. Only the method adopted to satisfy a building need should vary as to the procurement option chosen, reflecting a client's nature. The changing of consumers' lives from fundamentals of life expectancy to lifestyle choice, increased expectations of quality of life and quality of working life, increased demands on products and facilities and a paradigm of instantaneous gratification in construction clients' own markets lets them to expect in turn the construction industry to respond in a like manner in delivering construction projects. Although no building or facility can be completed instantaneously, clients are now demanding that contractors and designers compress overall design and construction programmes so that they can, in turn, compete more effectively in their own markets[165].

The efficiency and productivity of resources employed by clients needs to keep up with the general rise in competitive pressure for it to survive in an industry. Thus, the resources available to concentrate on an activity which is not usually a significant part in a client's value chain, such as the design and procurement of a construction project, need to be kept as low as possible, but, at the same time, have to ensure that construction needs are met on the most favourable terms and conditions available.

Therefore, there are two broad categories of construction markets. One, in which the client is experienced and has ongoing construction needs and the other, where a client is inexperienced with only a need for a one-off project specific item. The first type of

---

[165] Kubal, Miller and Worth, 2000, p. 11.

client finds himself in a unique position with a high degree of potential power over his suppliers, which is a useful position to be in when conducting any business related negotiation. He has the opportunity to organise his negotiation and selection processes professionally and is able to limit the number of suppliers who are awarded contracts in such a way that a group of preferred suppliers are created. A long-term relationship of such a nature can offer both parties benefits, a so called win-win situation if certain rules are obeyed, as shall be described later in section 5.1 when analysing the contractor's position.

The inexperienced client with only a one-off or casual need for building services, but whose project is just as important to him as it is for the experienced client, if not even more so, as it occupies much more of his resources and often represents a source of major disruption before it enhances competitive performance, faces a greater range of choice than ever before in deciding the most appropriate route of procurement. The procurement path he will choose all too often depends on past experience, however limited, and advice is given by trusted persons or organisations he regularly frequents. As was shown, in most cases this still results in the traditional method of procurement led by an architectural consultant without much thought of appropriateness. It has been stated by the chairman of UK's Confederation of Construction Clients (CCC) that half of all construction spending is by one-off and occasional clients and 80 % of that procurement is via an architect "met on the golf course". In his opinion, the chances of the traditional way of doing things bringing success are very low[166]. However, other methods are making inroads into the construction market.

The large number of inexperienced, one-off clients as well as the predominantly location based nature of construction are probably the two most significant factors accounting for the large fragmentation of the construction industry, particularly at the lower end of project size. Whereas at the upper end, the greater concentration of large and experienced clients seems to create a degree of intense competition among large contractors the world over, who are struggling if dependent on low-margin competitively bid projects and not being able to differentiate in any meaningful way.

---

[166] BAA, 2001, p. 50.

Whether a client is experienced and a frequent buyer or inexperienced and only a casual buyer of construction services, the approach and methodology of choosing a procurement route for a building or facility and the factors impinging upon it is described in detail in the chapter three.

It remains for a final comment on the distinction between private and public clients to point out that a public client, despite essentially being a frequent purchaser of construction services, is very much constrained by politically motivated procedural concepts in most countries, which practically turn the public sector into behaving as it were an inexperienced client and a casual purchaser of construction services. This comes as a result of having to decide each project on its own isolated merits of competitively bid lowest price and being forced to ignore any other factors[167] of capability which can normally be attributed to an experienced and intelligent client.

2.3.3 Analysis of contractors' behaviour and the consequences

As important[168] and experienced clients become ever more powerful and examine their supply chains with their new found market strength more carefully for potential improvements, it is not a surprise that construction, although not hitherto a particularly significant part of a client's value chain, has come under scrutiny as core activities have already received appropriate attention and additional sources of competitive strength are sought. The construction industry is not only being invited to do what it does better, it is being asked to join with its major clients in doing things entirely differently. Some clients have been clear about what they want, demanding that contractors they appoint build on time, on budget and to standards of quality which best meet their needs and they want shorten the design and construction supply chain, so that it can be better managed and become more flexible. They have realised that it is the large number of contractual interfaces and the confusion over responsibility for project delivery as a result that are very often the generating point of disputes and the principal cause of

---

[167] Factors that do get taken into account usually concern legal aspects, such as registration with the appropriate organisations, proof of an up to date tax record, correct employment practices and adherence to collective bargaining agreements.

[168] A client is referred to as important, when in the market for a series of substantial projects, sometimes anywhere in the world and exemplified by large multinational corporations.

delays and spiralling costs.

However, clients are generally dissatisfied with the performance of the construction industry, where the chief executive of the CCC in the UK criticised architects, quantity surveyors and contractors for offering clients poor advice[169]. It was stated that they needed to consider the priorities of its clients if its reputation was to improve and needed more integration between the client, the contractor and the supply chain. Architects and other consultants as well as contractors were criticised for not properly considering the client's business needs or that of the end users[170].

Such a statement highlights the situation that contractors are faced with having to satisfy increasingly demanding clients in circumstances that are characterised by stiffening competition, cyclical demand patterns, governmental interference and technological changes, all evidence of an increasingly dynamic and complex environment in which to operate. How is a contractor to satisfy a client's demands for a better quality product, in shorter time and for lower costs? Certainly not by performing its classical role of simply following a consultant's design, instructions and bowing under his management regime. To be able to serve a client satisfactorily, some contractors have set about overcoming this problem and have sought a direct relationship with a client as early as possible in the development of a project, in order to understand what the client needs and to go about as a team to satisfy these requirements while offering a single source of responsibility over the entire process. Such an approach to the procurement of a construction project is known as Design and Build and is described in much further detail in chapter 3.3 and subsequent chapters. Such a producer-led approach to procurement enables a contractor to lead the process from the front and offers the opportunity for optimising the entire supply chain contingent upon the needs of the client, including design, management, construction and, if necessary, the management and maintenance of the building or facility as well, as described in detail in chapter 4.3 . It requires a client to be willing to adopt such an approach to procurement of his building needs, assumes a considerable degree of trust in a contractor's ability to deliver

---

[169] see also comments in section 3.4.1.
[170] Lamont, 2001 b).

and considerable skill and capability on the part of the contractor to fulfil his obligations to the client's satisfaction. Not surprisingly, strong resistance is encountered by contractors from consultants, who suddenly find themselves in a supplier or subcontractor relationship. Although, some more enlightened consultants have seen the advantages and operate in a consortium or partnership role with the lead contractor.

The Design and Build method in combination with a long-term relationship with clients can generate a number of benefits to both client and contractor. This symbiosis is known as "partnering", which is defined by the United States Construction Industry Institute[171] as "A long term commitment between two or more organisations for the purpose of achieving specific objectives by maximising the effectiveness of each participant's resources. This requires changing traditional relationships to a shared culture without regard to organisational boundaries. The relationship is based on trust, dedication to common goals and understanding each other's individual expectations and values. Expected benefits include improved efficiency and cost-effectiveness, increased opportunity for innovation and continuous improvement of quality products and services." It has been said that partnering can be seen as a condition precedent to sound integration of project teams, however, it is important to recognise that it does not eliminate the need to structure the project organisation effectively and have rules of conduct in place[172], nor is it appropriate under all circumstances[173].

The situation just described is of real benefit only for clients with rolling construction programmes and possessing a structured and regular process spend for similar types of construction products and services[174] requiring a long term commitment[175]. The majority of construction clients, however, are in the market for only one-off and project specific items, that are often bespoke to the needs of individual clients and have to be built under varying conditions. Thus, the concept of partnering which is often referred to as the cure for all ills is really only suitable for a certain client, albeit an important

---

[171] Walker, 1996, p. 118.
[172] Ibid. p. 118.
[173] see chapter 5 for more information.
[174] CIOB, 1999, p. 31.
[175] Seely, 1997, p. 528.

one, especially for large contractors. Still, the concept of Design and Build, whether normally negotiated or with competition, is a means for contractors to become involved early in the process and they are thus placed in a position to shape it to suit their and ultimately their clients' needs. Taking this development to its logical conclusion it results in the client handing over all responsibility for a building or facility to the contractor, who is now more of a provider than a mere contractor and guarantees a specified level of performance over a substantial period of time in return for regular performance related payments over the period of the agreement. Such an agreement with a public sector client is usually termed a concession.

The advantages that a long-term relationship and commitment brings to the contractor are certainty of income, based on a regular flow of work of either new build or facilities management instilling an incentive for process optimisation, and, as a result of improved efficiency, a reduction of cost with savings shared between contractor and client. Once in a position to provide full life-cycle services and known to be expert in certain project categories (e.g. prisons, hospitals, schools) then the contractor has successfully differentiated himself from the majority of competitors, always remembering that constant change and continuous improvement are a prerequisite for ongoing competitive strength[176].

It is readily apparent from the above, that such a development calls for an altogether different set of skills and expertise than usually associated with a general contractor. The question of what actually constitutes the value chain of a general contractor arises and has been answered by a number of authors[177]. All tend to agree that a high incidence of knowledge based advantages reside in a project team in combination with both low and high order factors such as input of materials, capital, equipment, labour, which all require constant upgrading and improvement to provide a sustainable advantage and higher order factors to follow through into the creation of reputation and expertise in the execution of particular project technologies. This clearly is no longer feasible in a fast changing world of increasing technological development and

---

[176] For more information about competitive advantage, refer to: Porter; 1985.
[177] e.g. Porter, 1990; Klemmer, 1998.

specialisation coupled with variable demand, so that more successful general contractors have sought to concentrate on the management and co-ordination of construction projects. The aim of some is to provide a "cradle to grave" service for construction, maintenance and operational needs on a global scale, following the client across the globe.

This type of specialisation of general contractors into management and co-ordination of the entire construction process across the globe demands the lead position in the procurement process and is not usually welcomed by either consultants or suppliers, especially potential subcontractors, who prefer to remain autonomous and maintain direct relations with a client. This serves to protect an advantageous position, able to influence the client, maintain control over the cash-flow and generally enjoy greater flexibility and protection against outside competition. Another feature that protects some markets, as already pointed out, is that not all construction projects lend themselves to a method of world-wide partnership with single point responsibility for the entire process for the very existence of the inexperienced and casual client of construction services. Nevertheless, as larger contractors can develop a reputation for a quality product, enjoying some recognition in markets of many buyers such as housing, they may penetrate what have been to this day in most parts of the world geographically protected markets.

A contractor solely responsible for the delivery of a construction product nationally or internationally cannot rely purely on internal, directly employed specialists and labour or directly owned plant for a number of reasons[178]. For one, with the exception of office based designers and specialists, all personnel and plant need to be geographically flexible on relatively short notice for a limited duration, they need to be experienced and skilled in a number of trades and increasingly specialised techniques of fixing and installation. For another, on account of the project based nature of construction the demand for these resources is extremely variable, even in times of general economic growth. As noted when describing the increasing specialisation of construction, the growth of subcontracting has both been pushed by and has supported the changing

---

[178] see also section 5.3.2.

nature of general contractors, to the extent that in some countries they no longer employ direct labour or directly own plant and instead rely on subcontracting and plant hire. In fact, it is the ability of a main contractor to handle, co-ordinate and control subcontractors which is a decisive factor in maintaining a durable competitive position and deliver a quality product on time and on competitive terms to the satisfaction of the client. Chapter six will describe the multitude of factors that have to be managed in a general contractor / subcontractor relationship, which is of utmost importance for the wellbeing of both, despite the common perception that in the past and today a large number of such relationships are characterised by ill-feeling and distrust.

Major contractors are often placed in a position where economies of scale can matter in the supply of certain building materials or components, although, preferably, this is a matter handled more suitably by the relevant subcontractor, maintaining clear lines of responsibility[179]. But where a main contractor is the recipient of certain materials or components for a number of projects and involves different subcontractors none of which exhibiting the same kind of purchasing leverage, then it is worthwhile for a main contractor to utilise the potential and negotiate preferred terms and conditions with a supplier, thus enhancing his competitive position further[180].

Traditionally, architects have normally been solely involved in architecture and builders in building with very little overlap, if any. The various contributors have a tendency to focus upon and be concerned only with their own specialisms and are unable to perceive and respond to the problems of others[181]. Among members of a professional body, such as architects, sentience[182] has been found strongest if it confers upon its members the right to engage in professional relations with clients in which task and sentient boundaries coincide. There is a specific danger that when both direct relations with clients and coincidence of boundaries of sentient and task groups occur in that it may produce a group that becomes committed to a particular way of doing things. In the long

---

[179] A work package to be carried out by a subcontractor is best awarded as a whole including all matters such as detailed design, plant, labour and materials.

[180] see also sections 2.1.5 and 6.2.

[181] Walker, 1996, p. 109.

[182] A sentient group is one to which individuals are prepared to commit themselves and on which they depend for support.

run such a group is likely to inhibit change and behave as though its objective had become the defence of an obsolescent method of working. This view appears to have some significance for the construction process[183] and there have been many pleas over the years for the boundaries between the professions of the building industry to be broken down, as exemplified by the advice given at a recent strategic forum of the CCC in the UK in November 2001[184]. Construction clients (referring to casual, inexperienced clients) were advised not to approach architects or other industry professionals when they need a new building but, instead, seek out disinterested parties advice at a very early stage from an independent construction adviser[185]. Conventionally, the architect both designed and managed. Now, increasingly, a project manager is appointed to advice and manage the process on behalf of an inexperienced client, or who lacks the necessary resources to manage the process himself. Whatever the case, the manager's fundamental activity is integration, especially of design with construction. Some contractors, in meeting clients' desire to provide a much more efficient construction service, have engaged in new, and not so new, activities of contractor involvement in the design team, with Design and Build not only allowing an input of construction knowledge to the design but also potential for an equally important benefit in terms of integrating the contributors to the project. Its appeal to clients arises particularly form the single point responsibility which simplifies the manner in which the client interacts with the project team.

This process, where the contractor is to lead the entire design and construction process, places him at odds with the architectural profession, who are not content to forego their traditional role of client's representative and lead manager of the construction process and simply become a supplier to a Design and Build contractor. Architects do appear to support a traditional arm's-length orientation, while contractors prefer to integrate not only in terms of the organisational structure but especially in terms of compensation paid[186]. Consequently, contractors in a Design and Build position and extending the concept to BOT style projects, who want to manage the construction process from

---

[183] Ibid. 1996, p. 110.
[184] Fairs, 2001.
[185] See also comments regarding impartial advice in section 3.4.1.
[186] Puddicombe, 1997.

inception to completion and beyond, taking on substantial responsibility in return for performance related payments, are increasingly in direct competition with design consultants when it comes to establish trusted, first point relationships with clients[187].

Theoretically, to reduce differentiation to a minimum and have maximum integration, clients would develop their projects using a team of specialist skills as employees within their own organisation (in-house) including the construction phase using directly employed labour. In such a situation the likelihood of conflicting objectives among the contributors could be reduced. However, this type of arrangement is hopelessly uneconomic over the long term for any type of client for all the reasons already described. No single organisation can manage to be at the forefront of developments and achieve highest levels of efficiency and productivity in all fields simultaneously, particularly for a process as diverse as construction.

After all, it is the aim of a modern style main contractor to offer an integrated service, to be the first point of contact for a client in need of construction services and preferably maintain a long term relationship extending to a series of buildings and in some cases including full support over the life-cycle of a building, negotiated on terms favourable to both parties. Such a service offers scope for both parties to benefit in terms of reduced transaction costs, improved performance through learning curve effects and increasing specialisation in project technologies by bringing stability to the relationship. Undoubtedly, some criteria, such as bench marks and market testing or a select group of preferred contractors, to continually monitor performance and to ensure competitiveness of the service that a client receives will be considered[188].

Less experienced clients, who are only casual buyers of construction products, may well be looking for an integrated service from inception to completion of a building by a well known and reputable Design and Build contractor, especially if coupled to guarantees for fitness of purpose, price, time and backed by a performance bond. Perhaps such a relationship needs the comfort of an experienced advisor, probably best served by a

---

[187] See chapter 5.1 regarding working relationships.
[188] See chapter 6.3 for behaviour and control of main contractors.

professional project manager, who should be in a position to offer disinterested advice. A contractor to deliver a service as just described has to concentrate his efforts on integrating the full supply chain, including design, with technological expertise, management skills and business acumen. He has to make best use of preferred modalities where appropriate and be expert in handling subcontractors and suppliers as befits the needs of the project. This requires considerably more than simply locating the cheapest subcontractor or supplier and locking him in with seemingly fail-proof but inequitable and one sided often illegal contracts, but demands a long-term and intelligent approach in order to both capture and maintain the favour of a range of clients and a competitive position in a market driven by increasing competition. How this will be done the following chapters will describe in detail.

3.   Overview of Construction Procurement Types

3.1   Introduction

In an ideal world the client should have brought forward the initial need for the project in a co-ordinated and controlled way from within its organisation. The client should not have too rigid ideas at this stage of how to go about procuring his "ideal" building, but should have taken all useful advice from within its organisation before bringing forward for the project team's advice the strategies that it believes will fulfil the objectives.

It would be advantageous if the person who has co-ordinated the client's work and brought forward the strategies could form the client's component of the managing system when the project team is installed in whatever shape that may be, depending on the client's and project's specific criteria[189].

For any client to receive maximum benefit from its decision to procure a new building it is important to remember that the organisational issue, which incorporates the way in which people are organised and managed in the process of building procurement, is the most important element to decide upon at an early a stage as possible. This is to ensure that whatever techniques and tools are used and however well qualified people are it will be of no avail if they are applied within an inappropriate organisational structure. The question that now comes to mind is, of course, how to ensure that the organisational structure created to fulfil one's building need is indeed the most suitable ?

There are in the Anglo-American hemisphere any number of studies and authors that have described alternative ways of procuring a building with its corresponding organisational traits and have established in some cases fairly complicated systems to arrive at the "appropriate" procurement method for a particular client and its project specific criteria. However, in all of the models there is always a need to input specific characteristics of client, project and possibly procurement process features that are subjective based on the point of view of the user. Therefore, however more complicated the model has become it does not necessarily ensure that the final result is any better. Any model put forward needs at its heart an in-built relationship between client

---

[189] Walker, 1996, p. 146.

priorities and procurement method features that has been put together by knowledgeable people.

At times a particular method of procurement is seen as the ultimate answer by many "experts" in the industry and is claimed to constitute a "best practice approach". The emergence of a "new approach" is often heralded as a panacea for all previous problems. At first, there may be a number of notable successes, as the latest system is used under conditions for which it was originally intended. Then, as the new approach is more widely adopted and is used less and less appropriately it becomes only a matter of time before it becomes discredited[190].

Then again, there is a number of authors that argue that there is no such thing as a best practice method of procurement, only better practice. There simply cannot be a method of procurement that is always likely to be the appropriate way to achieve success under all circumstances. If there is one thing one can be certain off, it is the fact that technological and competitive circumstances do not remain the same. What can be done, however, is not to discuss a single approach as if it was the only way to success, but to consider it's appropriateness. This implies that one must choose wisely from amongst the range of potential procurement methods and corresponding organisational structures possible under the specific circumstances which affect them. It is not a fixed best practice method of procurement that must be sought, rather it is a recognition of what is appropriate and realistic. Sometimes it is claimed that procurement is significantly more complex and variable than construction academics and practitioners would like to have it and the variability is such that it is virtually impossible to classify procurement by any sort of rational positivist approach[191]. If then, in order to get the job done, a mixture of procurement types is thought necessary a bewildering array of construction contracts arises not without substantial transaction costs being incurred. Particularly the not so powerful or experienced client would be severely disadvantaged.

---

[190] Cox and Townsend, 1998, p. 32.
[191] Tookey, Murray, Hardcastle and Langford, 2001, p. 28.

It is only understandable if large and experienced clients wish to introduce their own onerous and perhaps one-sided contracts, thus exploiting their strong economic position. It should then not come as a surprise if disputes, claims and litigation are the result.

While the selection of an appropriate procurement system is an essential part of the building process, it is unlikely that an "ideal" procurement system that satisfies all criteria will be available to ensure success. A compromise solution is more likely, with particular strengths in key criteria areas, but also with weaknesses in certain areas of which clients and contractors need to be aware of[192].

In this respect the following chapters are going to differentiate at first between procurement systems and what can be described as "preferential modalities" that are not synonymous with particular procurement systems, but are generic types of good practice to all procurement activities[193]. Subsequently, an overview of procurement methods is presented, followed by a brief description of procurement path selection that will summarise the wealth of thought available on this topic and will produce what can be described as a guide to procurement selection. A general selection framework combines the methods of procurement available with the ideas of selection to offer a mechanism of impartial advice for further consideration. Finally, a brief summary of what constitutes standard contracts in the United States, the United Kingdom and internationally will conclude this chapter.

3.2   Differentiating between procurement systems and generic procurement techniques

As already pointed out, a distinction must be made between what is generally described by procurement system, the process by which the client seeks to satisfy his building requirement, characterised by a particular organisational form, distribution of responsibility, tasks and risk allocation, and what are generic types of best practice to all

---

[192] Ambrose and Tucker, 2001.
[193] Tookey, Murray, Hardcastle and Langford, 2001, p.22.

procurement activities[194]. They include, for example: supply chain management, lean production, investment in information technology and partnering[195].

These are aspects of management techniques or tasks which will aim at process improvements within procurement systems in the construction industry. This concerns the flow of materials, information and work processes and improvements in this area are made through innovations more often than not of the incremental refinement type rather than as a breakthrough leap, at least in the construction industry[196]. The overriding aim of many process improvements within procurement systems is to integrate the project processes of construction across key participants as much as possible.

How process improvements in the form of preferential modalities can be incorporated in any given procurement method is, of course, very much dependent on the organisational structure of the project team as instigated by the client or his advisor. There can be no hard and fast rules for the integration of the client and the project team, as, indeed, how the project team is put together. If one believes that once an appropriate procurement route has been selected much of the procurement process inevitably follows regardless of the consequences[197], then it is of paramount importance for the client to ensure that the mechanism selected is the result of an analysis of its own organisational structure, its needs and priorities and the circumstances of the project[198].

3.3   Types of procurement systems

3.3.1   Classification of procurement types

One of the consequences of procurement systems is to affect both organisational structure and process management of the project. The system of procurement will largely dictate whether the project is **designer-led** (i.e. architect or civil engineer in traditional systems), **project co-ordinator-led** (i.e. management type contract) or **producer-led** (i.e. contractor in Design and Build)[199]. Thus, the various procurement

---

[194] See also section 6.3.4 regarding early involvement tools.
[195] Ibid. p. 22.
[196] Ambrose and Tucker, 2001.
[197] Ibid.
[198] Walker, 1996, p. 199.
[199] Tookey, Murray, Hardcastle and Langford, 2001.

options available reflect fundamental differences in the allocation of risk and responsibility to match the characteristics of different projects and client needs. Selection, therefore, must be given strategic consideration[200], but should be undertaken at that stage where consideration is given to the appointment of designer, manager or producer[201].

Other terms to describe the three different groups of procurement system are shown in the table below.

Group[202]		
1	**2**	**3**
design-led traditional conventional design-bid-build	**project co-ordinator-led** management fee construction	**producer-led** design and construction design and build[203] design-build single source package deal BOT

Table 4: Terminology for different procurement groups

While it is very convenient to classify procurement systems and their characteristics of organisation and distribution of risks in such an orderly fashion, it must be pointed out that the situation is very complex at times, particularly where the client has introduced his own in-house documents or variations to a standard contract document. Some authors have gone as far as stating that it is virtually impossible to classify procurement systems by way of a rational positivist approach[204]. It must be realised that divisions between procurement types are being blurred by developing practice. Having said that, much effort, pain and ultimately costs can be saved if clients or their advisors, based on the characteristics of organisation and risk allocation wished for, decide on one of the procurement types that the majority of authors and practitioners agree upon. They have the additional benefit of being able to rely on tried and tested standard contract formats, which will be described in chapter 3.6.

---

[200] See chapter 3.4 for additional information regarding selection processes.
[201] CIOB, 1999.
[202] e.g. CIOB, 1999; Cox and Townsend, 1998; Halpin and Woodhead, 1998; Newcombe, 2001; Pilcher, 1997; Seely, 1997; Smith, 1995; Tookey, Murray, Hardcastle and Langford, 2001; Walker, 1996.
[203] The term „Design and Build" will often be used to denote „producer-led" procurement systems, as it is the most common type.
[204] Tookey, Murray, Hardcastle and Langford, 2001.

3.3.2   Presentation of procurement types

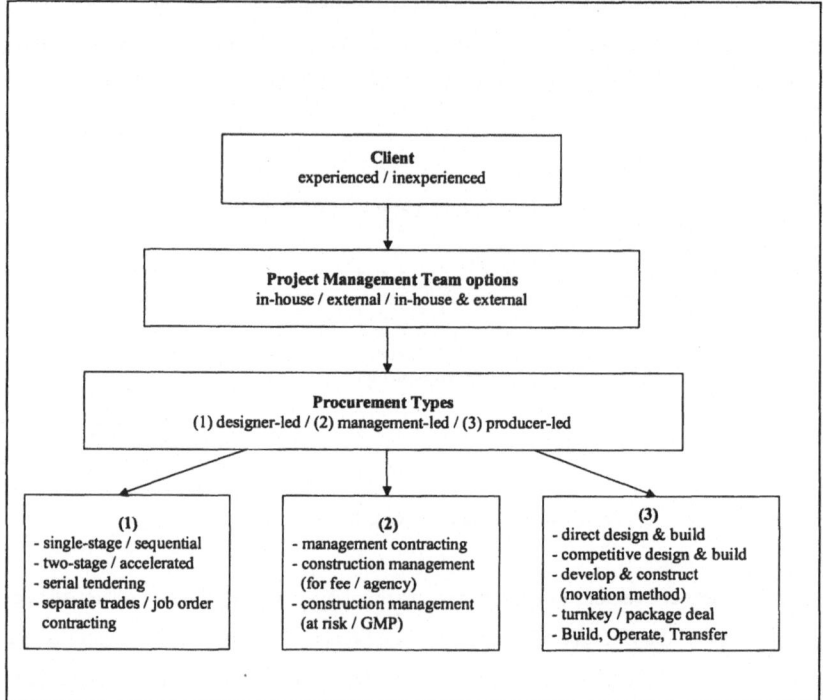

Figure 9: Procurement types

This chart presents in a very concise way all procurement methods known[205]. It is probably sufficiently diverse to allow the classification of all procurement systems of

---

[205] e.g. Cox and Townsend, 1998; Seely, 1997; Smith, 1995; Walker, 1996; Gralla, 2001, for a German text on this topic.

construction wherever they occur[206]. For example, Germany's convention of classifying construction procuring systems differs, but can be located in the above chart, where "traditional" relates to separate trades contracting (Gewerkevergabe) and general contracting (Generalunternhemer) equates to traditional, single stage in the chart. The type of "Generalübernehmer" would translate into Design and Build or possibly package deal in the chart. The overriding principle behind this type of classification is based upon the allocation of responsibility over the project development process, primarily design. It will always be architects and engineers doing the design, however, there are differences in terms of authority. Group (1) is governed by designers, traditionally architects or civil engineers, depending on the type of project. Group (2) is management-led, meaning that either a ontractor or consultant is appointed to manage design as well as the construction process, but does not himself undertake any construction. Group (3) is led by the producer, who is generally the contractor managing the construction of the project. In this case he will be responsible for all matters concerning the project, including design and carrying the risk of cost and time. In extreme circumstances such as BOT projects it would be probably more correct to speak of a provider or promoter, meaning that the promoter has adopted the role of client in respect of the project and will seek on his behalf the best procurement route for the facility required. Certainly, the promoter is as a rule a very experienced client and hence

---

[206] An important distinction must be made between procurement types and payment methods. While certain payment methods are particularly suitable for a given procurement type, they are essentially interchangeable. For example, both traditional or Design and Build can either be based on admeasurement (unit rate), lump sum or target cost (GMP) terms of payment.

Lump sum / stipulated sum: where the contract price is based on a single tendered price for the whole works. Payment can be in stages, according to a defined stage of progress.

Admeasurement / unit-price/rate: where the contract price is based on a Bill of Quantities or schedule of rates. Payment is usually at monthly intervals and is derived from measuring quantities of work completed and applying rates in the tender, or new rates negotiated from tender rates.

Cost reimbursable / negotiated: where the payment is based on actual cost plus a specified fee for overhead and profit. It involves open book accounting and the contractor is reimbursed by periodic progress payments. Different types of fee structures are applied and include: cost & percentage of work, cost & fixed fee, cost & fixed fee & profit-sharing clause, cost & sliding fee.

Target cost: A reimbursable type of payment, whereby a client and contractor agree at the start a probable (or target) cost for a then uncertain scope of work. Any difference between the actual cost and the target cost at completion is shared in a way that is defined by the incentive mechanism. In some cases the target value is used to define a **Guaranteed Maximum Price (GMP)**, which the contractor guarantees will not be exceeded. In this situation, any overrun of the GMP must be shouldered by the contractor. It may be defined as the target plus some fraction of the target value, e.g. if the target is 100 million, a GMP of 105 million might be agreed. Seely, 1997; Smith, 1995; Halpin and Woodhead, 1998.

directly involved in the construction process. The following chapters will provide more detail on the various procurement types.

3.3.3 Designer-led tendering

In the UK as well as in the US the system of traditional tendering was based on the rigid separation of the design and construction activities from the beginning of the nineteenth century[207]. The client appoints a team of consultants with the architect as team leader, responsible for both design and management of the project. Following a feasibility study and the development of the detailed design, where the design team prepares all drawings, specification and a detailed bill of quantities, the process of tendering for the selection of a suitable contractor takes place. In many cases each consultant, including architect, engineers and quantity surveyor[208] will be independent of the other contributors. Yet, the contributors will need to be interdependent in terms of the project. The contractor is likely to be in contact with a large number of suppliers of materials of all kinds and subcontractors for carrying out the works and equipment installations[209]. Indeed, it has been stated that "it is currently universal practice in the UK for the specialised trade skills to be provided by independent trades contractors, or even self-employed individuals. It is virtually unknown for a general building contractor to provide building skills from internal, directly employed resources"[210]. Some of the suppliers and / or subcontractors may be nominated by the client or on his behalf by one of the consultants. Such subcontractors will normally be selected after submission of their tender to the client and the contractor is then instructed to enter into a contract with the nominated subcontractor on terms that are specified by the client or the consultant. Other subcontractors, those arranged by the contractor, are known as domestic subcontractors and are usually subject to the approval of the architectural or engineering consultant[211].

---

[207] Cox and Townsend, 1998, p. 34.

[208] The QS will give advice on a range of matters relating to the cost of the work as well as preparing some of the contract documents and measuring the work for valuation and variation purposes together with the preparation of the final account. For more information about the role of the quantity surveyor refer to: Seely, 1997; Winter, 2000.

[209] Pilcher, 1997, p. 200.

[210] CIOB, 1999, p. 29.

[211] Pilcher, 1997, p. 28.

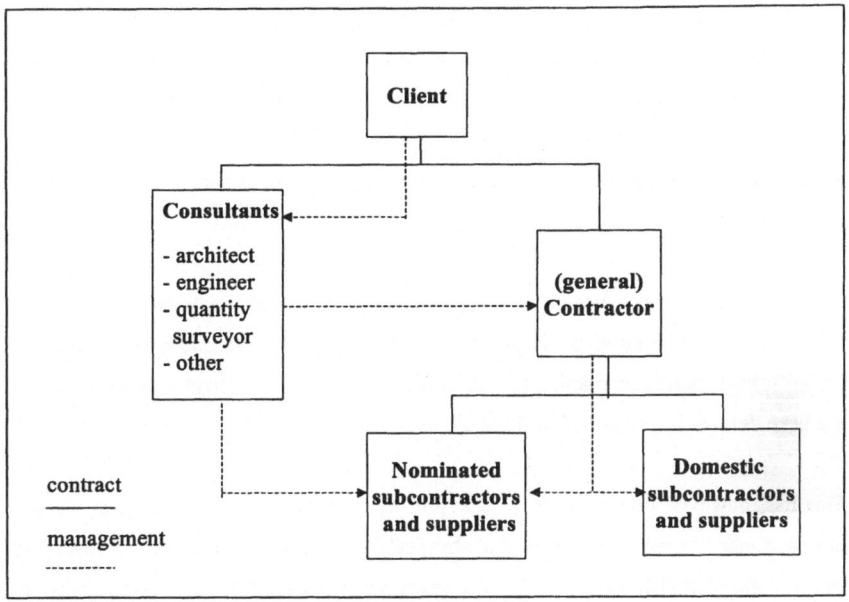

Figure 10: Traditional procurement structure

The advantage of the traditional system is seen in its ability to "test the market" and on the one hand demonstrates so called "public accountability" for public funds where the client is a public authority and on the other hand "good value for money" for commercial / private clients[212]. The lowest tender is regarded to be the most suitable, as all other criteria are laid down in great detail, thus allowing no variation in quality and fitness for purpose, since all drawings, specification and bill of quantities are available. In fact, especially the availability of a full and accurate bull of quantities in the contract documents is regarded as a well recognised practice with the following perceived advantages:

- Bills avoid the need for all tendering contractors to measure the quantities themselves before preparing an estimate. This saves on wasteful duplication of effort and an increase in the contractor's overheads which eventually has to be passed on to the client.

---

[212] Newcombe, 2001.

- Bills prepared in accordance with a recognised standard method ensure that an adequate description of the works in a recognised format is given to all tendering contractors and therefore all tender on the same basis. The absence of bills leads to greater variability, increased risk in estimating and consequently more disputes.

- The detailed breakdown of the contract sum permits proper financial management of the contract.

In addition, it is very easy to compare and evaluate all tenders based on the same information since price alone is the variable factor, as long as one assumes that all design information is available, it is all correct and serves the client's purpose, which has been defined in great detail right at the outset, in all respects.

This has, however, proved to be a fallacy in all to many cases over the past and the lowest tender has not always been the cheapest for a number of reasons, but basically because the overwhelming significance based on the tender process as the basis for assessing value and the selection of the contractor has caused adversarialism, directed attention away from the total acquisition cost, life-cycle costs and value and has perpetuated the fragmentation of the construction industry[213].

Such a structure produces a high level of differentiation between all contributors, which demands a high level of integration. Unfortunately, traditional tendering requires that the contractor, who is to construct the project, cannot be introduced at the design stage. The problem of integration is further complicated by the fact that the managing system is not differentiated form the operating system. The architect is attempting to fulfil dual roles, one is the operating system of design, the other is the management of the project. In this type of situation there is a high potential for someone not to be able to exercise objectivity in decision making. Whoever is in the position is placed under severe pressure by being required to undertake tasks that frequently are often incompatible skills: design and management[214]. This criticism does not necessarily apply to all projects, but the more complex a project becomes the more likely it does. A traditional

---

[213] Cox and Townsned, 1998, preface.
[214] Walker, 1996, p. 120, Rösel, 1994, p. 102, Sommer, 2000, p. 20.

procurement system, thus:

- restricts access of other contributors to the client.
- inhibits the client from approaching other contributors for client advice.
- has no one solely in a project management role.
- causes integration within the design team to be difficult to achieve, as can be integration of the design team with the client.

The situation as described above can arise whenever the first contributor to be appointed is an engineer, a quantity surveyor or other consultant and has advised the client on the appointment of the other consultants. The first appointed contributor assumes project management responsibilities alongside professional functions leading to a potential lack of objectivity[215].

Attempts to overcome some of the limitations in the traditional single-stage approach, mainly the problem that arise from the total separation of the design and construction process and allow the contractor to be involved to some degree in the design stage have led to the emergence of two-stage or accelerated traditional tendering in the UK during the 1960's. This is achieved by a higher level of integration and exposing the design team to the management discipline and expertise of the contractor. The design team establishes a notional bills when the design has reached a suitable stage of completion. Selected main contractors are then asked to tender rates against the approximate quantities contained in the bill and may be requested to submit their proposals for the management of the construction operation and any suggested design changes / improvements. The successful tenderer is then involved in the further development of the design as a member of the project team. The full bill is prepared when the design is fully developed with rates transferred directly from the notional bill or negotiated if there are substantial differences to be considered. The greatest benefit are the opportunity to involve specialist subcontractors in the design and the opportunity to accelerate the construction programme[216].

---

[215] Walker, 1996, p. 201.
[216] Cox and Townsend, 1998, p. 35; Walker, 1996, p. 208.

Two-stage tendering represents a trade-off between integration of the contractor into the design team and some potential for accelerating the construction programme against a conventional approach to competition. However, designers are often sceptical about the contribution a contractor may make to the design of a project and is not very likely to happen easily, where the decision to integrate has not been made by the designers themselves[217].

Other variations to the traditional approach include continuity contracts, primarily developed to reduce transaction costs and save time[218]:

- ad-hoc/negotiated tendering, where there is a negotiation of rates for a second project based on those form the first contract.

- term tendering, where the contractor is appointed for a fixed period of time (often between one and two years) and is reimbursed in accordance with a comprehensive unit schedule of rates.

- serial tendering, where a client effectively batches a series of similar projects together on the basis of a notional bill in order to introduce greater economies of scale and some time saving by avoiding repeated tendering.

The US equivalent method of procurement (design / bid / build) is similarly based on a competitive process, where the contractor with the lowest total bid is responsible for the construction works of the whole contract. However, it differs from the UK approach as often there is only fairly basic design and construction documentation available and a major effort on the part of the general contractor together with his subcontractors is required to produce the design detailing necessary[219]. A perfect set of plans and specifications is rarely, if ever, produced by design consultants and even if that were to be the case one contractor's interpretation of these plans and specifications may vary from that of another contractor. When differences in contract interpretation result in additional costs, which is bound to happen when the quality of the contract documentation is poor, then some form of dispute or claim will arise[220].

---

[217] Ibid, p. 208.
[218] Cox and Townsend, 1998, p. 36.
[219] Dielschneider, 2000, p. 23.
[220] Levey, 1999, p. 25.

One of the consequences arising out of the difficulties with the traditional tendering method, which includes the need for all contract documentation to be totally complete prior inviting bids resulting in a sequentiality of design, is its extension of the total design – build time frame. The shortening of time by designing and constructing in parallel is not possible hence the adaptation of the traditional method by way of two-stage tendering.

Since "traditional" contracting in other, non Anglo-American, parts of the world is usually the term used to describe separate trades contracting, it needs to be included as a form of procurement method. It is a generic title for an approach were there is no main contractor appointed for the project on a lump-sum basis but instead a member of the client's organisation or fee earning member of the project team, usually the architect, organise trade contractors to undertake the work and are responsible for running the site and directing the activities of the trade contractors. Trade contractors are directly contracted to the client on either a lump-sum or unit-rate, sometimes even hourly-rate competitively bid contract. In effect the site architect replaces the main contractor's site agent and provides the site with direct and constant design supervision. The client's involvement on site is usually higher than on conventional, general contractor run sites. It is claimed that communication is as direct as possible from client to architect to tradesman, that the human element is all important and the client's interest is best served by people committed primarily to the client's project rather than their profession or trade[221]. However, it does not incorporate construction expertise as such in the design stage, but relies upon the ability of the architect in this respect and in respect of running the whole construction process with all its difficulties.

3.3.4 Management-led tendering

It is important at this stage to point out the difference between the concept of project / program management and construction management systems, which often causes confusion even among experienced practitioners and authors[222]. It is perhaps the Americans which have the clearest understanding of the difference, carefully

---

[221] Walker, 1996, p. 211.
[222] As witnessed in the introduction of: Watson and Speak, 2001.

delineating between what they call program management and construction management. The project manager is always both acting on behalf of, and representing the client and its leadership function is essentially about managing people[223]. The title "project manager" should have a reserved meaning in the construction industry as projects are executed for clients and as the title means managing the project as a whole, then it should refer to managing the project for the client, that is the specific and overarching objective of the project manager must be achievement of the client's objectives[224]. The project manager must seek to resolve conflict in the process in the interest of the client and hence must act as a professional consultant without entrepreneurial interest in the project.

The title does not always have this reserved meaning in practice and this leads to confusion. It is particularly persons performing management activities within the construction process, such as construction management, contract management and design management that are often designated project manager. These activities, however, do not necessarily have the client's interest as their main concern but owe allegiance to the business objectives of their own organisation. For example, the so called project manager of a management contractor, particularly when at risk, is distinctly different from the client's project manager as his focus is not solely on the client's objective[225].

Whilst every project has to be managed, it must also be recognised that a separate or external project manager may not be required and that many clients have their own in-house project management capacity, especially if seen in combination with one of the enhanced construction services in the form of either producer or management-led procurement systems to be described hereafter[226]. Of course, project management need not only occur in combination with construction management but can be applied with any of the procurement types described. It is even likely, that if a professional, external project manager is engaged the more direct routes of either design and build or separate

---

[223] CIOB, 1999, p. 4.
[224] refer also to: Seely, 1997, p. 333.
[225] Walker, 1996, p. 7.
[226] Ibid, p. 152.

trades will be chosen, since the services offered by management-led procurement routes lend themselves to more experienced clients who have chosen not to employ an external project manager.

A common feature of a variety of management systems[227] in construction is, that a client enters into a contract with an external construction management organisation, which should be integrated in the management and co-ordination of design and construction of the proposed works. All physical construction is undertaken by subcontractors or works contractors selected either in competition or negotiation. There are two basic types which again have themselves some major differences, usually in the allocation of risk that a contractor is willing to bear and is shown in the figure below[228].

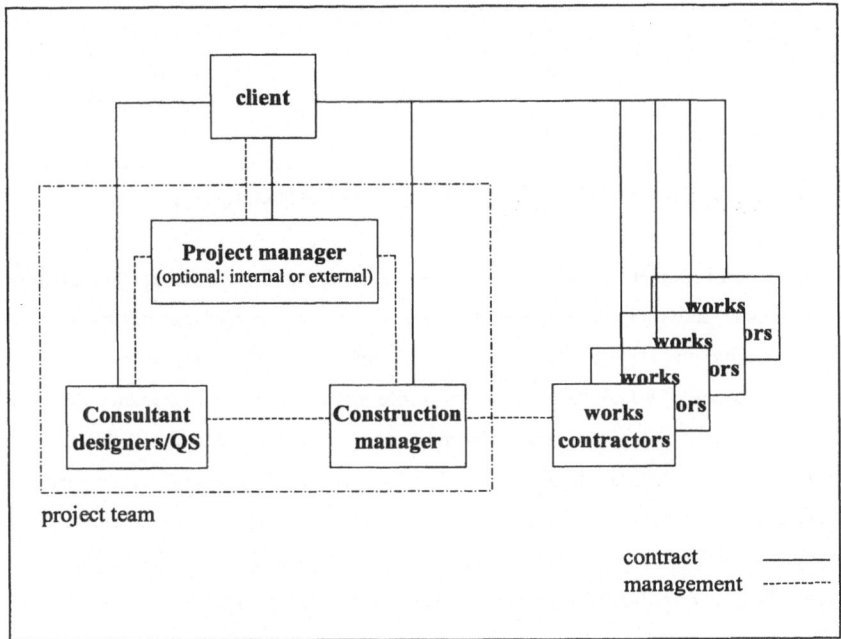

Figure 11: Management-led tendering / Construction Management

---

[227] Which have evolved in the United States in the 1960's, e.g.: Cox and Townsend, 1998, p. 39; Watson and Speak, 2001, and are estimated to account for approximately 20 % of all projects in that country; see: Levey, 1999, p. 27.

[228] Halpin and Woodhead, 1998, p. 74; Pilcher, 1997, pp. 29-30; Seely, 1997, p. 98.

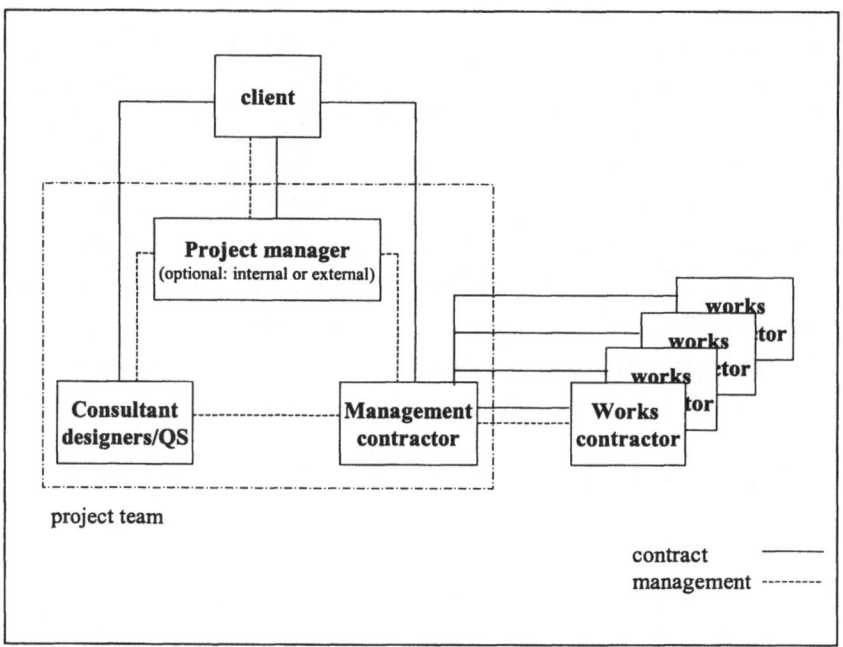

Figure 12: Management-led tendering / Management Contracting

Construction management or management contracting can be defined as "the group of management activities related to a construction program, carried out during the pre-design, design and construction phase that contribute to the control of time and cost in the construction of a facility"[229]. It serves a role that can substantially reduce the work load of an external project manager or may even make him redundant if the client is sufficiently experienced. It is this situation that causes most confusion in people's understanding of what constitutes project management. Matters are made worse if the client has conferred the project management duties to one of the members of the project team who may be either the designer or the construction manager.

Construction Management requires that the specialist works contractors are contracted to the client directly, leaving the construction manager, as a member of the consultancy

---

[229] Halpin and Woodhead, 1998, p. 73.

team under the direction of the project manager, to concentrate on the organisation and management of the construction operations[230].

Management Contracting or as it is referred to in the US as "Construction Management at risk (AGC)[231], Construction Management – Contractor (AIA)[232]" is similarly to Construction Management a method where the construction is carried out by specialist works contractors, but in this case they are actually sub-contractors who are contracted directly to the management contractor on terms approved by the project manager.

Essentially, the difference between Management Contracting (MC) and Construction Management (CM) is defined by the degree of involvement with physical construction works where, for example, a MC can be expected to provide central site facilities (site set-up, preliminaries) and the degree of integration with the project team (i.e. designers and cost consultants)is for a MC usually not as on an equal basis as it is the case for a CM. The MC is normally brought in at a later stage, primarily on the basis to deal with a large number of different works contractors and not as much as a construction advisor[233]. CM recognises the role of management as an explicit professional function separate from contracting[234]. The CM is appointed in a similar manner to the other professional consultants with similar liability to the client. This procedure avoids some of the drawbacks of MC, which can prove to be more confrontational and expensive and carry a greater degree of risk for the client, works contractors and management contractors[235]. The circumstances for which the management-led type of procurement is adopted are the same for both MC and CM, but for reasons just explained even more so for CM, and are suitable for conditions, where[236]:

- large, complex projects are undertaken.
- there is a greater requirement for flexibility on design changes then conventional systems will allow.

---

[230] CIOB, 1999, p. 29.
[231] AGC – Associated General Contractors of America.
[232] AIA – American Institute of Architects.
[233] e.g. Cox and Townsend, 1998, p. 39; Seely, 1997, p. 95.
[234] Cox and Townsend, 1998, p. 40.
[235] Seely, 1997, p. 97.
[236] Cox and Townsend, 1998, p. 39.

- there is a need for an early start for the construction phase, a need for early completion but the design is insufficiently developed.
- there is a need to consider particular construction methods during the design phase.the client and designers have insufficient management resources, and
- a large number of different contractors are regionally based, resulting in many interfaces for co-ordination.

The management-led approach is most suited to various combinations of the following[237], where:

- the client is familiar with construction processes and techniques and knows some or all of the professional team (client without the services of an external PM).
- the project is technically complex, involving diverse techniques and subsystems.
- the client requires an early start on site and a fast-track approach.
- the client needs to retain the right to make variations to requirements as the project proceeds.
- the nature of the project is such that it is realistic to separate professional responsibility for its design from professional responsibility for its management.
- the client wishes to retain the flexibility to use competitive tendering and / or negotiation for procuring separate elements of construction.
- the cost to the client needs to be competitive, but the control of cost in terms of seeking value for money is more important than simply securing the least possible cost, and
- the client wishes a less adversarial form of contract.

It can be said that because the construction manager is not at risk in the way that a building contractor would be in a conventional method, and he has no means of increasing his profit margin, his attitude to the project will be similar to that of the professional team. For example, he will be concerned with keeping costs of the works within the project budget price, which he would have had a say in, reporting to the client or project manager on possible extras and who is dealing with subcontractors

---

[237] CIOB, 1999, p. 96; Cox and Townsend, 1998, p. 40.

(works contractors) in regard to such matters as claims for loss and expense and the settlement of accounts[238]. While CM is a positive approach to the integration of construction expertise into the design process it does not serve the client in obtaining greater cost certainty from the outset or necessarily provide a single source of responsibility for his construction procurement needs. The disadvantages can, therefore, be noted as: cost and time uncertainty and no recourse in this respect, higher financial risk for the client and greater engagement in the details of the construction process[239].

So far, management-led procurement types have been described in terms of remuneration of the MC or CM occurring on the basis of a fee, not necessarily related to the performance. Equitable performance measurement is often difficult[240]. A variation on the topic of CM has developed for this reason, which is known as CM at risk, where the construction manager, rather than simply working for an agreed fee, will in addition guarantee a maximum price for the project in return for providing services similar to that of the CM for fee[241].

The CM at risk is now deemed to have an incentive to manage the construction works in such a way as to stay within the GMP and benefit from a share in the savings as agreed with the client or project manager. On the other hand, if the completed project exceeds the GMP, any cost overruns are absorbed by the CM, thereby reducing the CM's profit, or indeed causing him a loss[242]. As the GMP price is agreed after design work has sufficiently developed in order to estimate total costs reasonably accurately (akin to agreeing a price in Design and Build) he in effect becomes now a general contractor, which could give rise to a clash of interests if he seeks to maximise profit and minimise risk by introducing safe prices with subcontractors and unreasonable safety margins in longer construction programmes[243]. It would further appear that the CM at risk has

---

[238] Seely, 1997, p. 95.

[239] Watson and Speak, 2001.

[240] CIOB, 1999, p. 118.

[241] This is not to be confused with the AGC understanding of „CM at risk", where not only a guaranteed price but also the works contractors are contractually linked to the CM, just as in MC, and a further modification has led to the absence of a GMP (AGC 565) making it in effect a standard MC as practised in the UK.

[242] Levey, 1999, p. 32.

[243] Dielschneider, 2000, pp. 28-29.

somewhat vested interests in ensuring that costs remain less than the GMP and some authors claim that this limits his objectivity. A CM at risk may view a works contractor's claim for extra payment or extension of time differently or he is less than co-operative when it comes to design changes by the owner, unless he has settled on additional compensation first. The observation of many practitioners in the US has found that, in the absence of a highly qualified owner's project manager, the approach of CM at risk has a capacity to create disputes, claims and costs. It has been claimed that as a result it is not very popular in the United States at present. Other, more effective means to provide an incentive for CM agents to improve their performance, apart from growing reputation and obtaining referrals, is to allow a premium if the project is brought in below budget and time or reducing fees if the project does not perform as planned[244].

3.3.5   Producer-led tendering

This chapter is limited to describing the mechanics of producer-led procurement methods. Since the intention of this work is to concentrate on just this type of procurement method the advantages and disadvantages of using this type of approach to construction procurement will be discussed in more detail in subsequent chapters, after a guide to the procurement selection process and a general selection framework have been presented.

Single source systems are a group of procurement systems that enable clients to employ one firm only to take the responsibility for the complete delivery of their construction needs[245]. They are arrangements that do not separate design and construction as the one firm offers the total package of design and construction. At least from the client's point of view it becomes the responsibility of one organisation, which usually is a contractor, for delivering the required building and associated services in accordance with defined standards and conditions[246].

---

[244] Levey, 1999, p. 28-29.
[245] Cox and Townsend, 1998, p. 32.
[246] CIOB, 1999, p.28; Cox and Townsend, 1998, p. 37; Halpin and Woodhead, 1998, p. 72; Kubal, Miller and Worth, 2000, p. 34; Levy, 1999, p. 34; Ling, Khee and Lim., 2001; Pilcher, 1997, p. 28; Seely, 1997, p. 97; Smith, 1995, p. 152; Walker, 1996, p. 210.

There is traditional Design and Build[247] and then there are varying degrees of involvement of the contractor with the management of the design process and involvement with the actual construction of the project. Beside Design and Build there are "package deals" and "Turn-key" construction and in recent times a number of other variants have emerged, including "Build, Operate, Transfer (BOT)". They are effectively similar in concept, the difference is in the balance of responsibility between client and contractor.

The selection of a Design and Build contractor should be based on a brief of the employer's requirements. Although the contractor assumes the overall responsibility for project delivery, the client may appoint an independent advisor to help in developing the brief and to monitor quality and costs. This is especially the case if the client does not have the necessary in-house skills to arrange for tenders for the work to be submitted and then for their evaluation and the selection of a suitable contractor.

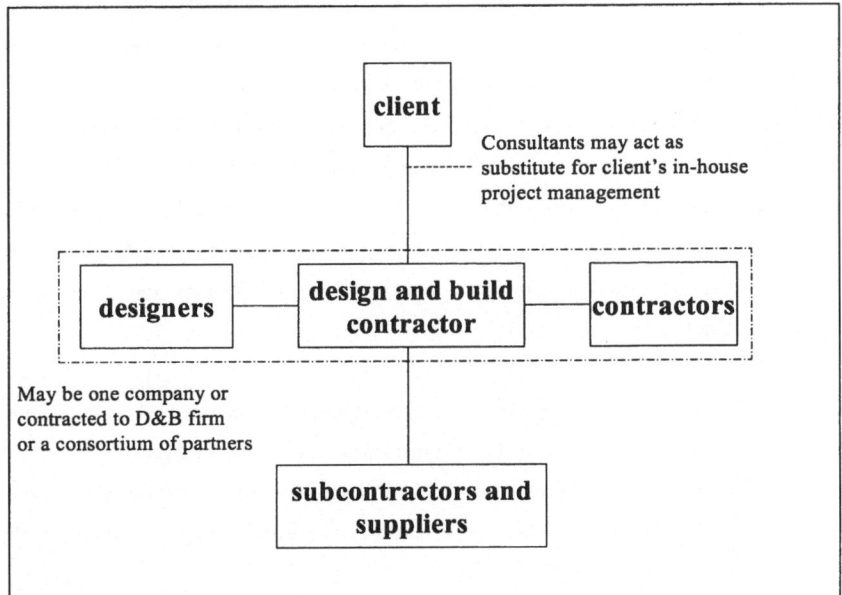

Figure 13: Design and Build

It was in the 1970's that large construction firms began to offer this type of service in order to provide the client with a single source for project delivery of industrial construction of a complex nature that had tight time requirements such as petrochemical plant, power plants, etc. .Usually, only firms with large design and construction capabilities were able to provide design and build services and projects built as such were often referred to as "Turn-key" projects[248]. In the past, the use of Design and Build contracts has become much more common in the building sector in the US and UK and elsewhere in the world[249]. Since most building contractors do not have an in-house design capability, lead contractors typically form a team or consortium of designers and specialist contractors who work together to meet the needs of the client. Contractors nowadays generally prefer to subcontract design due to the greater access to a wider range of design skills provided by this approach and the reduction in risk associated with not having designers on their staff[250].

Generally, the variation of design and build can be classified as follows:

- direct (traditional) design and build, where the contractor is appointed often on appraisal and negotiation, but no price competition.
- competition, where there is price and design proposal competition between several firms on the basis of a conceptual design proposed by a consultant.
- develop and construct, where the design is partially completed by the employer's designers, typically after 30 % to 40 % of the project design is completed[251], before contractors are asked to complete and guarantee the design and price in a competitive tender.

The latter is a method very much favoured by a number of clients, but who introduce a further amendment, where the client's designers who have developed the project to the point of appointment of the Design and Build contractor are passed to the contractor for the completion of the project. This is referred to as "novation Design and Build" with

---

[248] Halpin and Woodhead, 1998, p. 72.
[249] Dielschneider, 2000; Halpin and Woodhead, 1998, p. 72; Ling, Khee and Lim, 2001.
[250] Walker, 1996, p. 212.
[251] Halpin and Woodhead, 1998, p. 73.

the contract between client and designer novated to the contractor, who then also bears all risks of design and construction associated with the project[252].

"Turn-key" or "package deal" can be seen as a concept that carries the method of Design and Build further. The contractor still provides the design and construction but will in addition provide construction finance for the project. When the project has been completed and the "key has been turned over" to the client, full payment is made[253]. Alternatively, all the client / promoter would have to do would be to literally "turn a key in the door" and the project would be operational, since the contractor was responsible for deign, procurement, engineering, finance and commissioning[254].

"BOT"[255] is a further development of the design and build type of procurement, where the contractor effectively becomes the client in respect of the project and has become its "promoter" and the client is now the "principal" receiving the benefit of the project. The contractor, in addition to the role of turn-key contractor, not only finances the project during the construction stage, but finances it, operates it and maintains the building / facility over a period of time, thus, generating sufficient income to provide a commercial return. This is a concession type of contract granted by the client (principal) for a stipulated period of time, usually somewhere in the range of 20 to 30 years depending on the type of project and on occasion considerably longer, subject to meeting all contractual obligations contained in the concession agreement.

The analysis of advantages and disadvantages of these types of producer-led procurement systems will be discussed in depth and greater detail in chapter four and section 4.5.3 summarises the preferred application of contractor–led procurement.

---

[252] Walker, 1996, p. 212.
[253] Levey, 1999, p. 34.
[254] Smith, 1995, p. 241.
[255] Other typical acronyms used to describe types of concession contracts include: BOOT – Build, Own, Operate, Transfer or DBFM – Design, Build, Finance, Operate. For more acronyms and additional information regarding Privately Financed Concession Contracts refer to: Merna and Smith, 1996a);b).

3.4   A guide to the procurement selection process

It is possible to begin with a substantial list of authors and organisations that have thought about, described and have put forward best practice methods for selecting the appropriate procurement type. However, this would not serve meeting the objective of investigating the appropriateness of contractor-led procurement and a guide, therefore, which sums up most of what has been written and provides a simple yet useful starting point for a general selection framework of procurement types.

3.4.1   Problems encountered during selection

A major concern of clients is that they get little impartial advice about whether they need to build and the best way to go ahead. Questions that describe their concern can be[256]:

- Do I need to build or is there some other alternative?
- What choices do I have about ways of building ?
- How do I select the right procurement path for me ?

The most significant problems potentially arising during the procurement process were identified as changing requirements (28 %), design team problems (25 %) and communication (18 %) and, interestingly, the majority of the respondents saw the solution to their problems as a change in the procurement system[257]. At the same time, determining the abilities of a procurement path for all the various permutations and combinations of client and project features is not an easy task and it is then understandable why clients and building professionals by and large resort to the procurement system with which they are familiar, regardless of the appropriateness for the project and client[258]. It has been shown that most clients consistently use the system with which they are most familiar or rely on professional advice. Unfortunately, those most likely to offer clients advice, i.e. architects, engineers and quantity surveyors, were also found to be the least inclined within the construction profession to suggest or seek

---

[256] Newcombe, 2001.
[257] Ambrose and Tucker, 2001.
[258] Ibid.

change. This suggests that many clients are ignorant of their procurement system option.

Indeed, Latham in his report[259] has gone as far as to suggest that clients, who are unable to undertake their own project strategy or need definition in-house, are well advised to retain some external expert, but not initially in the form of a Project Manager. Such a consultant is there to help the client decide if the project is necessary. If a professional advisor has been retained in the expectation of being lead consultant for the project, it will only compromise that individual in advising the client whether or not the project is needed, and if it is, that it could be done with a small scheme requiring no further or limited consultant advice. Any client who wants external advice over project strategy and need definition should only consider an advisor on the express understanding that the role will terminate once the advice has been formulated on whether or not to proceed.

3.4.2   Organisational features of projects

A consequence of procurement selection is to affect both the organisational structure and process management system for the project in mind. Features of project organisation to be considered are[260]:

- The relationship of the project team to the client organisation and the client's influence upon the critical decisions.
- The degree of interdependency of tasks and people generated by the project organisational structure.
- The degree of differentiation present within the operating system, which should be reduced to a minimum. The level to which it can be reduced will be constrained by the nature of the project.
- The level of integration provided by the managing system and the complexity of the managing system itself. Over elaboration can lead to severe differentiation within the managing system, which should have the operability to match its integrative effect to the degree of differentiation present in the project.

---

[259] Latham, 1994.
[260] Walker, 1996, p. 198.

In short, there are three major components to the organisation structure of projects:

- The client / project team integrative mechanism.
- The organisation of the design team.
- The integration of the construction team into the process.

Whatever procurement system is chosen by or on behalf of the client, it should be the result of an analysis of the client's organisational structure, the client's needs and experience and the complexity of the project. There certainly is not a single, ideal method of procurement which satisfies all clients under all circumstances, thus there can be no hard and fast rules for the integration of the client and the remainder of the project team. However, there are approaches that offer some guidance as how to go about in selecting the most appropriate procurement path for a particular client and his specific construction needs may it be producer led or otherwise.

3.4.3 Management approaches for determining selection criteria

Walker[261] describes a set of functions which are to be performed by the client or his project manager, either from within the client's organisation or externally, that are necessary to define a client's selection criteria for the appropriate procurement system. These are:

- Establishment of the client's objective and priorities, based on its broader organisational and project objectives. This should allow the development of a brief for the project.
- Design of the project organisation structure, which should be based upon critical decision points that have to be made in the process and take into account the relationships of the contributors among each other and to the critical decisions.
- Identification of the way in which the client is integrated into the project, which will arise form the design of the organisation referred to above. It is important that the client meshes with the project team, responds to the need to integrate with the project team and is aware of effective communications.
- Advice on the selection and appointment of the contributors to the project and the establishment of their terms of reference, where it is a matter of experience of the client whether he seeks advice or even leaves this entirely up to an external

---

[261] Walker, 1996, pp. 147.

consultant. Perhaps the most difficult decision that the client will have to make is whom to appoint to manage the project and how to integrate with that project manager from within his own organisation, depending upon the extent to which the client wishes to retain power of approval.

- Translation of the client's objectives into a brief for the (potential) project team and its transmission, which involves the establishment of user needs, the budget, cost and investment plans. It is at this stage that fundamental misinterpretations can occur and opportunities for economics are overlooked, which then become enshrined within the development of the project. The client has to ensure that its objectives are clearly transmitted to potential contributors and is understood by them. There is a perpetual danger that it will be misinterpreted and result in contributors once selected to pull into different directions.

Another method, by Cox and Townsend[262], describes a four-fold approach that "better-practice" companies adopt. By segmenting their total schedule of construction requirements they can begin to devise a portfolio of procurement strategies. Essentially a distinction is made between experienced clients (regular construction spend) and inexperienced clients (one-off construction spend) as well as considering the supply market conditions for a particular product or service as either difficult or not. The approach, as the name implies, breaks down into four stages:

(1)   Segmentation thinking, where a client thinks carefully about the nature of the supply chain it is involved in and begin to differentiate and align the internal operational activities in such a way as to focus on internal and external customer needs and wants within each supply chain.

(2)   Critical supply chain asset management, which refers to a way of thinking that differentiates between types of supply chain that the company is embedded in and the structural properties of each type of supply chain. These are critical, complimentary and residual supply chain assets and to exist successfully in its chosen supply chain position a company has to buy on an operational level many products and services that are either highly complimentary or of residual importance to the primary activity that the company is focused on. Any product

---

[262] Cox and Townsend, 1998, pp. 322.

or service purchased is in fact always processed in a supply chain. There is always a high or low degree of vertical integration and a variety of competitive market structures in place within any supply chain. It is these structural properties, rather than the finished product or service purchased, that a client must manage operationally based on its own specific circumstances.

(3)   Analysis of supply market conditions, which is aimed at increasing procurement competencies and can only occur if a client has a solid grasp of the structural properties of the supply chain from which it buys. Understanding the existing structure of power within the product or service supply chain and the capacity for the company to change the balance of power in such a way that an improvement in cost, quality and time can be achieved to be more efficient and effective than its competitors is the objective.

(4)   Achieving a "strategic and operational alignment" of the relational competencies that flow through their primary and support supply chains is what tends to characterise a successful company. "Relational competence thinking" is the capacity to link all the operational and strategic supply questions about how to source external products and services appropriately.

In essence, the approach described here is about aligning strategic and operational practices with a portfolio of relationship types in order to achieve a desired corporate outcome. This way of thinking methodologically is referred to as "critical asset and relational competence analysis"[263].

3.4.4   Client criteria

Having described management approaches for adoption by a client to determine its needs in terms of building procurement, the process will have resulted in a number of criteria, which will have to be met by the type of procurement path chosen. A client cannot be expected to know the strength and weaknesses of the various procurement systems, but he will know what he expects from whatever procurement path is selected.

---

[263] See also section 6.2.2 for procurement classification and strategies from the perspective of a main contractor.

The objectives of a client are influenced by factors internal and external to the client and traditionally clients prioritised the basic criteria of time, cost and quality. However, whilst fundamentally correct, it is simplistic and the themes need to be developed. There exist quite a number of studies, research reports and works on this particular topic of client's priorities and the selection of procurement method in the Anglo-American world, but the results by most authors tend to repeat each other and have adopted similar approaches aimed at a decision support system[264]. Core objectives of clients that literature has indicated can be summarised as[265]: highest realistic quality, lowest realistic cost, minimum realistic time into service, high prestige for the building (within affordability parameters) and minimum conflict during the process. The following questions indicate the range of matters which the client will generally need to consider to make the most appropriate choice of procurement path[266]:

- Design input: does the client want to influence the design and, if so, to what extend?
- Client control: how involved does the client wish to be in the management of the project?
- Cost certainty: what level of cost certainty does the client require before signing the contract and on completion of the project?
- Risk taking: is the client prepared to accept the risk by direct management or does he wish to transfer it to another party?
- Flexibility: to what extent is the client's brief likely to be changed during the execution of the project?
- Market conditions: how are market conditions likely to change during the course of the project with possible consequences for design or construction?
- Programme security: how crucial is the final completion date?
- Value for money: does the client want to contribute to and take benefit from value management and value engineering, and how will any resulting savings be shared?
- and additionally: which other generic management tools mentioned earlier does the client want to introduce?

---

[264] Ambrose and Tucker, 2001; Tookey, Murray, Hardcastle and Langford 2001; Wong, Holt and Cooper., 2001.
[265] Tookey, Murray, Hardcastle and Langford, 2001.
[266] Ambrose and Tucker, 2001; Seely, 1997, pp. 66.

A discerning and recognised set of procurement criteria was established by NEDO[267] in 1985 and refined by the Business Round Table in 1995[268], and they are: timing, controllable variation, complexity, quality level, price certainty, competition, management, accountability and risk avoidance.

3.4.5 Project criteria

Whilst attempts have been made to view project criteria separately from overall client needs[269] with some good supporting arguments, as it is generally better to allow for interaction between client needs and project characteristics, it complicates matters further and increases the permutations and combinations of features used to decide upon which particular procurement type best accommodates all these factors.

Main categories of project characteristics are:
- Level of complexity of the project, either design and / or construction.
- Repetitive nature of the process, where low rise residential, warehouses and car parks are examples of works of a repetitive nature.
- Risk associated with the construction, such as technical, economic or political risks.
- Scale, where very large projects require high levels of control to keep the whole project on track.

Rather than trying to separate overall client needs from strictly project specific criteria, it is advisable to allow for improved practical application and merge both aspects under a single heading of procurement criteria. It is difficult anyhow to always separate a client's needs from strictly project specific criteria and ultimately it is the client's overall needs that define his decision making and not project specific criteria on their own.

The NEDO approach, as refined by the Business Round Table[270] and expanded further by the author, adopts such an approach, where for ease of implementation both the

---

[267] National Economic Development Board.
[268] Ibid. pp. 69.
[269] e.g. Ambrose and Tucker, 2001.
[270] Seely, 1997, pp. 70.

client's needs and project specific criteria are merged into one set of procurement criteria that need to be addressed. For any "decision support system" the question of accuracy and, therefore, validity is a moot point. Results generated from a number of studies of decision support systems have demonstrated significant variability[271]. There appears not to be a single "decision support system" in the area of procurement system selection that will under all circumstances deliver the right answer. Having said this, however, a simple but at the same time sensible approach can be a valuable first-step guide, with the result to be further analysed and tested against client's requirements and consultants' advice.

3.5 A general procurement selection model

The client, who from within its organisation has both decided upon the need for construction works and its objectives and has defined its criteria, should consider each procurement option to ensure that the managerial and contractual arrangements between itself and the rest of the project team are fully understood. The principal working methods of each route with their inherent advantages and disadvantages should be discussed and finally the most appropriate type has to be determined for the project. The following instructions refer to the selection model shown over the page and is intended as a primer for discussion with the principal advisor and should not be used as the sole basis for making a procurement decision.

The procurement criteria established so far and previously discussed reflect the priorities of the client and are listed on the left hand side. The procurement options or paths described earlier are shown along the top row with the right hand side reflecting the importance of the criteria as well as the ability of the procurement path to satisfy it. The multiple choice answers to each question outlined on the right hand side of the chart are considered and the answer which appears most relevant is identified and the appropriate dot on the chart is marked. When all the questions have been considered the number of marked dots in each column is totalled and the procurement path with the most marks should be worthy of further investigation.

---

[271] Tookey, Murray, Hardcastle and Langford, 2001.

Once again it must be stressed that procurement selection is not a science as there are multiple variables involved, often of a subjective nature. However, the main advantage of the selection model presented here is not necessarily to give the "correct" answer, but that it forces participants in the construction process to consider alternatives and agree on a reasoned case for a particular procurement type. In this sense it presents a system of independent and impartial advice to clients without the need to be knowledgeable about different procurement paths.

The procurement selection model cross-tabulates thirteen procurement types (grouped as **producer-led**: BOT, Turnkey/Package-Deal, Direct D & B, Competitive D & B, Dev. & Const. D & B; **management-led**: MC at risk, MC for fee, CM at risk, CM for fee; **designer-led**: separate trades, serial tenders, two-stage, single-stage) against the following selection criteria and priorities:

type	criteria	priority	options
A	price certainty	Do you need to have a firm price for as much of the procurement process as possible before you can	yes / budget only
B	timing	How important is early completion to the success of your project ?	crucial / important / not important
C	controllable variation	Do you foresee the need to alter the project in any-way once it has started on site ?	yes / some / no
D	complexity	Is your building of a high design or technical standard and can the project environment be described as dynamic, moderately so or not dynamic ?	yes / moderately / no
E	quality level	What level of quality (standard) do you seek in the design and workmanship of your project ?	basic / good / prestige
F	contractor input	How important is the ability to involve contractors' expertise at the design stage ?	important / not important
G	competition	Do you need to choose your construction team and/or work contractors by price competition ?	work contractors / wks. & const. mgt. teams / no
H	management	Can you manage many separate consultants and contractors, some, or do you want just one firm to be responsible after the briefing stage ?	many separate firms / some separate firms / one firm only
I	accountability	Do you want direct professional accountability to you from the designers and cost consultants ?	no / yes
J	risk avoidance	Do you want to pay someone to take the risk of cost and time slippage for you ?	no / share / yes
K	operation & maintenance	Do you want to pay someone to take the responsibility not only for designing and building, but for the operation and maintenance of your building as well ?	no / share / yes

(Bottom row: TOTALS)

Figure 14: General procurement selection model

The following table represents some results for hypothetical cases and their particular sets of client criteria[272].

case no.	type of client	client criteria	procurement path
(1)	industrial producer (factory)	- quality shell - short time to market - no experience of construction - little management involvement - moderately complex	1. package deal or BOT 2. direct D & B
(2)	up-market food retailer (supermarket)	- good quality building - experienced with in-house design team - time and cost certainty required - moderately complex	1. two-stage, MC for fee, MC at risk, dev. & const. D&B 2. single-stage
(3.1)	local authority "traditional set-up" (admin. offices)	- must demonstrate lowest costs by competition - moderately complex - accountability required from designers - flexibility of design	1. separate trades 2. single stage, CM for fee
(3.2)	local authority "concession type" (admin. offices)	- value for money demanded - moderately complex - accountability required from provider - certainty of price, time and quality	1. BOT 2. Turn-key, competitive D&B
(4)	financial institution (new headquarters)	- prestige building - complex - client wants to be involved - flexibility of design - accountability by designers - inexperienced client, but with external PM	1. CM for fee, MC for fee 2. two-stage, CM at risk
(5)	developer (offices)	- great price certainty required - moderate quality - experienced client - some management involvement - high level of risk avoidance - max. delegation of operation and maintenance	1. BOT 2. Turn-key

Table 5: Procurement paths suggested for further analysis

It must be noted that those aspects concerning producer-led procurement method are based largely on what can be described as traditional or conventional views of performance ability, where, for example, it is not considered particularly suitable for very complex or prestigious construction projects. This will be addressed in later chapters[273] and methods will be discussed that enhance the circumstances under which producer-led procurement options become an appropriate alternative. Nevertheless,

---

[272] Actual worked examples are found in the appendix.
[273] see chapters 4 and 5.

producer-led procurement options have already accounted for two out of five projects, or alternatively for three out of five projects.

3.6   Standard construction contracts

A construction contract is a binding document, enforceable in law, containing the conditions under which the construction of a facility takes place. It results from an undertaking made by one party to another, for a consideration, to construct works that are subject of the contract. The offer in construction is normally in the form of a tender and when full and complete agreement about the conditions and the consideration (usually payment) has been reached the acceptance can be formalised.

Traditionally, consultants for professional services will be required to be in contract with their employers, so that the parties to the construction contract are the contractor and the employer / client. The architect, quantity surveyor, engineers and other consultants are not parties to the construction contract. Each has their own terms of employment with the employer, usually on a standard form issued by the appropriate professional body, and at individually agreed fees[274]. With producer-led procurement, the consultants will typically not be contracted to the client, but to the contractor, however, are still outside the construction contract that exists between client and contractor.

An organisation that enters into a large number of contracts each year often evolves a standard set of conditions that establishes their procedures and applies them to all construction contracts. This set of provisions is normally referred to as general conditions. For those organisation that enter into contracts on a less frequent basis, or for all those organisations that wish to benefit from established, tried and tested construction contracts, professional and trade organisations publish standards that are commonly used in the industry and reflect the considerable variation in procurement approaches. The standard from of contract sets out to establish a series of relationships that apply to most types of building projects and enables people working on different types of projects to carry out their tasks in as standardised a form as practical. In setting

---

[274] Pilcher, 1997, pp. 30; Seely, 1997, pp. 18.

up these relationships the standard form covers the method of ordering work by the client, dealing with delays and default on both sides, arrangement of insurances and calculation of the final account together with stage payments and variations in cost[275].

It is always advisable, whenever possible, to use standard conditions of contract that have been agreed by representatives bodies of the construction industry, including the professional institutions and client representatives. Such conditions have normally been tried and tested in practice and modifications to them should not be made without the appropriate legal advice[276]. This is explained by the fact that the contract language embodied in the contract has been hammered out over the years from countless test cases and precedents in both claims and court actions. The wording has evolved to establish a firm and equitable balance of protection for all parties concerned. Another feature of standard forms of contract is that a family of contract documents has evolved, where not only the main contract between client and contractor, but also other contributors to the project are covered by separate contracts, which are tailored to suit the type of main contract. Since the major risks and responsibilities have already been efficiently allocated, the user of standard documents saves on considerable transaction costs. They have an industry accepted foundation for their transactions and no longer need to go through a protracted negotiation process for each transaction risk. Rather, they and their legal and insurance advisors may only need to revise transaction specific additions to and deletions from accepted standard document forms. The following will provide a brief overview of standard construction documents in the United States, United Kingdom and internationally[277].

3.6.1   Standard documents in the United States

Standard documents in the United States are prepared by the Associated General Contractors of America (AGC) and the American Institute of Architects (AIA). Both have a very similar approach in that a contracts documents committee composed of a large number of members, who are experienced practitioners in industry and the professions, law, insurance and other sectors from across the country, seek to provide

---

[275] Ibid. p.118.
[276] Pilcher, 1997, p. 36.
[277] See also 4.5.4 for a brief reference to the German standard contract.

and continually improve balanced documents for the construction industry. As an example of how extensive such a document family can become the "AGC 200" series shall be briefly presented here[278].

The AGC 200 series represents traditional contracting activities with the central document entitled: "Standard Form of Agreement and General Conditions Between Owner and Contractor (Where the Contract Price is a Lump Sum): AGC 200, 2000 edition)". It is intended to form an integrated agreed general conditions document between the owner and the contractor performing work on a lump sum basis. It is appropriate for use in competitive bid environments or in situations requiring a negotiated sum contract. The architect is contracted to the owner on a separate but compatible document, the AGC 240 document.

Other documents within the AGC 200 series are: forms for changes in work – AGC 202 and 203, performance bond – AGC 260, payment bond – AGC 261, bid bond – AGC 262, contractor's qualification statement for engineered construction – AGC 220, certificate of completion – AGC 260 and 261, application for payment – AGC 291 and 292 and schedule of values – AGC 293. More documents in this series include forms to cover variations on the lump sum, for example where the basis of payment is the cost of work with a fee for pre-construction services – AGC 230.

Other document families include the 400 series for use with Design – Build, the 500 series for Construction Management in either agency (for fee) or at risk format, the 600 series for subcontracting documents (where the AGC 650 is intended for use with AGC 200 and compatible with AIA 201 and the payment to the subcontractor is not conditional on the contractor receiving payment from the owner, as distinct from AGC 655 where the payment to the subcontractor is expressly conditioned on the contractor receiving payment from the owner) and the 800 series for program (project) management agreement and general conditions between owner and program manager.

---

[278] For more information the reader is referred to www.agc.org or www.e-architect.com. The AGC, established in 1918, is an organisation of qualified construction contractors and industry related companies. It currently comprises more than 100 chapters and 34,000 firms including 7,500 general contractors and 26,500 industry associates, who are subcontractors, speciality contractors, suppliers, equipment manufacturers and professional firms.

The contractual configuration of the latter document is of a "pure / agent program manager" not at risk, either with all design and construction contracts signed by the owner or the program manager signing the contracts as the agent of the owner. The program manager can be seen as replacing the owner's facilities staff and may oversee a project delivery accomplished under a variety of methods (i.e. Design, Bid, Build or Design – Build) for each discrete project or site.

While similarities exist between the two document families, the AGC specifically states that AGC and AIA documents are as a general rule not compatible and should not be used together, since in almost every instance a document is intended to work only within its own respective document family.

3.6.2 Standard documents in the United Kingdom

While the modular approach to contract documentation as practised in the United States is not as developed in the United Kingdom, there still exist a number of standard documents having the same attributes as described above. The most widely used documents were traditionally the JCT Standard Form of Building Contract, intended primarily for competitively bid building contracts and in its original form on a lump sum basis. The Joint Contracts Tribunal (JCT) over the years developed a number of other standard contracts including for use with Design and Build and Management Contracting. The Institute of Civil Engineers (ICE) Conditions of Contract are used primarily for civil engineering works (basis for the old FIDIC contract) and it has produced other standard contracts such as for Design and Build. The ICE did adopt a completely new approach to engineering contracts with the key objectives of namely flexibility, clarity and simplicity and the promotion of good management, when it created the New Engineering Contract (NEC) in the early 1990's[279]. It has been claimed, that the Engineering and Construction Contract as it is now known, can be used on any engineering, building or construction project in any country and on any scale. It comprises a core contract and six main options encompassing a conventional contract with activity schedule, conventional contract with bill of quantities, target

---

[279] Refer, for example, to Bennet, Baird: NEC and Partnering – The Contract to Building Winning Teams, Thomas Telford. 2001 and McInnis, Wilde: The New Engineering Contract – A Legal Commentary, Thomas Telford, 2001 for more information about the NEC contract.

contract with activity schedule, target contract with bill of quantities, cost reimbursable contract and management contract with a number of secondary options for use where necessary to allow the client to choose the version most appropriate for his needs. Latham in "Constructing the Team" suggested using the Engineering and Construction Contract because it is flexible enough to be used with all types of procurement strategy, although not without some alterations[280].

Consultants are employed on their own terms of employment by the client which, for example, are the Association of Consulting Engineers (ACE) (Conditions of Engagement", the Royal Institution of British Architects (RIBA) "Architect's Appointment" and the Project Management Association of the Royal Institution of Chartered Surveyors "Project Management Agreement & Conditions of Engagement" for project managers[281].

3.6.3 International standard documents

Prior to 1998, the Fédération Internationale Des Ingénieurs – Conseils (FIDIC)[282] published three forms of building and engineering contracts:

- for civil engineering works (known as the Red Book[283]),
- for electrical and mechanical works (known as the Yellow Book), and
- for Design and Build (known as the Orange Book).

The foreword of the 1987 edition of the Red Book stated that "the clauses of general application have been grouped together in this document and are referred to as Part 1" and contain the terms which are not expected to be changed. Part II includes the "conditions of particular application" which must be specially drafted to suit each individual contract, for which some example wording and guidance for drafting

---

[280] Hill, 2000, p. 6; Latham, 1994, pp. 36; Seely, 1997, p. 122.

[281] CIOB, 1999, p. 191; Pilcher, 1997, p. 32.

[282] International Federation of Consulting Engineers. FIDIC memberships number 56 countries from all parts of the globe, representing most of the independent consulting engineers in the world. FIDIC, 1994.

[283] The first and original edition of FIDIC started in 1977. The current edition is entitled „Conditions of Contract for Works of Civil Engineering Construction, Fourth Edition 1987, Reprinted 1992 with further amendments. There is an additional publication to be used in conjunction with the Red Book for subcontracts, which is the „Conditions of Contract of Subcontract for Works of Civil Engineering Construction".

purposes is given. It warns that users need to take care in avoiding errors in their drafting of each Part II, particularly where example wording is varied or additional clauses are included to avoid ambiguity with Part I or between the clauses in Part II, and tenderers have to be careful to read the tender documents thoroughly[284].

The new books for major works comprise FIDIC's four 1999 first editions:

- Short Form of Contract, which is recommended for building or engineering works of relatively small capital value and which may also be suitable for other relatively simple work or work of short duration.

- Conditions of Contract for Construction (the Construction BOOK or CONS), which are recommended for building or engineering works where most of the design is provided by the employer. However, the works may involve some contractor designed civil, mechanical, electrical and/or construction works.

- Conditions of Contract for Plant and Design – Build (the Plant & D-B Book or P & DB), which are recommended for the provision of electrical and/or mechanical plant and for the design and execution of buildings or engineering works. The scope of this Book thus embraces both old Yellow and Orange Books, for all types of contractor designed works.

- Conditions of Contract for EPC[285] / Turn-key projects (the EPC Book or EPCT), which may be suitable for the provision on a turnkey basis of a process or power plant, of a factory or similar facility, or of an infrastructure project or other type of development, where i) a high degree of certainty of fixed price and time is required and ii) the contractor takes total responsibility for the design and execution of the project. However, it has to be used as and where appropriate and with care and professionalism.

All new books for major works (excluding the Short Form) are published in three parts:

- General Contract
- Guidance for the Preparation of the Particular Conditions (GPPC), and
- Letter of Tender, Contract Agreements and Dispute Adjudication Arrangements.

---

[284] Booen, 2000.
[285] EPC = Engineer, Procure and Construct.

The basic concept underlying the structure of the three major new books is to provide maximum convenience for users, particularly those who prepare the tender documents. In order for them to choose from the alternative arrangements provided in the new books, they must posses a reasonable understanding of procurement and contractual procedures and anticipate possible events during the execution of the type of works involved. It is the FIDIC's intention for those who write tender documents to find it easier to concentrate on the particular procurement aspects of the project, rather than having to concentrate too much on typical provisions within the Particular Conditions only[286].

---

[286] Ibid.

4   Contractor-led Scenarios

4.1   Introduction

In chapter three the mechanism and the function of contractor-led procurement either in the form of Design and Build, Turn-key and BOT was explained. Now it is time to address contractor-led procurement in more depth, particularly from the point of view of its appropriateness for delivering clients' construction needs and where it is less so. It will draw on the discussion of earlier chapters, especially those on market trends and individual participants' objectives and behaviour, in order to establish what contractor-led models of construction procurement can offer to the client to satisfy his construction requirements.

Contractor-led procurement in the guise of Design and Build contracting can trace its beginnings back to the master builders of ancient and medieval times, where the master builder completed both the design and construction and self-performed all site activities. Under this contracting method there was clearly one single point of responsibility – the master builder. This was the case for the building of the pyramids, the building of Rome during the Roman empire, for the building of the great cathedrals during medieval times and was still the case in the 17th and 18th centuries[287]. Only with the realisation that time is money in more recent times has the need for speeding up construction and increasing mechanisation meant a separation of design and construction. Another aspect is the transfer of legal liability when moving away from Design and Build and self performing work to a separate construction and design contract, with the design obligation transferred to the client, who now must indemnify the general or trade contractor for design errors under the contract.

Industry clients have responded to the problem of strict design and construction separation by turning back the clock to the proven method of Design and Build, which places all liability on a single source under its basic and original form. The increasing demands for ever faster delivery of the construction product may wholly move the industry towards wider acceptance of the Design and Build method of contracting and may become the procurement method preferred by progressive firms that rapidly

---

[287] Kubal, Miller, Worth, 2000, pp. 18; Rösel, 1994, pp. 12.

respond to competitive pressures of the market place. Whether this is a viable proposition, appropriate under what circumstances, shall now be investigated.

In the United States to date, where Design and Build has started in the 1970's, the spread of Design and Build is such that it was estimated to account for one third of construction projects in the mid-1990's[288] and that it will rise further to 40 % by 2005, having stood at 15 % in 1990[289]. In the United Kingdom the market share for Design and Build projects was approximately 25 % in the mid 1990's[290]. A comparison of work being undertaken by different procurement methods in the UK is shown below[291].

work won during July – September 1998	no. of projects	no. of projects as a percentage %	total value (£ million)	total value as a percentage %
Construct. Mgmt.	29	11	495	26
Traditional	194	70	994	52
Design and Build	53	19	416	22
Totals	276	100	1905	100

Table 6: Comparison of work by different procurement methods

4.2   Organisational features of contractor-led procurement

Design and Build may potentially provide the most effective integration between the design and construction phases and can be even more all encompassing than traditional Design and Build when considering package deals and BOT arrangements. As the contractor accepts total responsibility for both the design and construction of the project the opportunity to provide effective integration of the processes is theoretically higher in Design and Build approaches than in more conventional systems[292]. A properly integrated Design and Build organisation can operate on a project team basis, with those possessing different but complementary skills getting to know and respect each other. There are distinct advantages in enabling the contractor to use his management skills and experience in the pre-construction period to ensure that design and performance are

---

[288] Ling, Khee and Lim, 2001.
[289] Dielschneider, 2000, p. 29.
[290] Ling, Khee and Li, 2001.
[291] Watson and Speak, 2001, p. 24.
[292] Walker, 1996, p. 211.

closely co-ordinated and better related to time and cost. Economy and efficiency should flow from the continuity of joint experience[293].

From a client's perspective the allocation of responsibility for the project among potential contributors is under Design and Build the simplest, whereby the project is managed by a single firm which appoints the consultants directly (i.e. are "subcontracted" to it) and the managing firm takes the responsibility for their work and hence the total project. Thus, this system has the advantage that differences or disputes between the design team or group and the construction team are matters internal to a single project organisation. Normally, the management of the Design and Build contractor is motivated to reconcile disputes or differences between design and construction in as timely and efficient a manner as possible, since, if such problems are not addressed, they can lead to significant losses and potential dismissal of the contractor for poor performance, including substantial claims for damages incurred by the client.

Some limitations apply to this principle under some of the variations to Design and Build, where, for example, the novation Design and Build method, in which the client's design consultants, who have developed the project to the point of appointment of the Design and Build contractor and are then passed to the contractor for the completion of the project, presents less opportunity for the contractor than the traditional direct Design and Build format. The perceived advantage of this variation from the client's perspective is to develop his requirements with a designer of his choice, but retain the certainty that Design and Build brings in respect of time and cost and overall responsibility of design is delegated to the Design and Build contractor after novation.

Naturally, the greater the responsibility accepted by the firm the greater the risk they are carrying[294]. Design and Build places more responsibility and liability on to the contractor than any other form of procurement[295]. Firms are unlikely to accept higher responsibility, and therefore risk, unless they have direct control over the contributors

---

[293] Seely, 1997, p. 100.
[294] Walker, 1996, p. 146.
[295] Hughes, Gray and Murdoch, 1997, p. 31; Levey, 1999, p. 240; Simm, 2000.

direct employment, alliance or subcontract, or a facility to bring an action against a contributor if one is brought against them by the client[296].

The organisational factors so far described apply largely to both types of Design and Build organisations, whether they are in-house or a consortium of lead contractors, designers and specialist contractors. Some difficulties of integration may occur if a contractor subcontracts all other consultants, however, an experienced Design and Build contractor is likely to work with a network of like minded contributors and can develop a "project group" most suited for a particular project.

A contractor-led procurement system is most suitable for applying generic procurement techniques such as partnering and supply chain management, since a single entity responsible for both design and construction can introduce optimal processes across the supply chain down to the smallest subcontractor or supplier[297].

What has been said of contractor-led procurement so far applies just as well to turnkey projects and BOT arrangements, which take the process even further with the delegation of responsibility for construction finance or even maintenance, operation and possibly the total service[298], but involves the contractor / provider / promoter to take on even greater responsibility and therefore additional risk. Furthermore, as the service provided goes beyond the simple provision of design and construction, additional contributors are necessarily involved, usually in a consortium type of organisation called a Special Purpose Vehicle (SPV), to provide finance, facility management and operational expertise.

A situation in which responsibility for the project rests with only one firm, or at least one firm for the overall management of design, construction and related aspects, such as finance, maintenance and operation, is likely to be attractive to clients. Informed clients are in a position to dictate the pattern they want for their project and in return the contributors expect to be appropriately recompensed for their risk.

---

[296] Walker, 1996, p. 146.
[297] See also chapter 6.2 and 6.3 for appropriate application.
[298] For example a power station, where the contract is for a specified quantity of power.

Some organisational aspects of contractor-led procurement, which are seen as a potential problem, are:

- Difficulties of integrating the project team with the client, where the client should have a clear conception of its objectives, but those of the Design and Build firm may at times conflict with those of the client and may result in some constraints being placed upon the client developing his requirements and variations possibly proving to be expensive.

- Design and Build firms having a tendency to be orientated towards construction activity, traditionally having a construction background, which may have a detrimental effect for the integration of design and a subsequent effect upon its quality.

- The emerging relationships possibly facing difficulties in sufficiently integrating the design and construction team in cases where the Design and Build contractor "subcontracts" design[299].

4.3   Positive features of contractor-led procurement

This chapter is to analyse the positive features of contractor-led procurement, such as Design and Build, Turn-key and BOT, under aspects of time, cost and quality and will refer to previous chapters of procurement type and the discussion on organisational features of this type of procurement.

4.3.1   Positive features of contractor-led procurement in respect of time

Studies both in the United States and the United Kingdom have shown, that Design and Build projects not only experience the fastest construction activity and therefore shorter construction periods, but also offer the shortest overall project delivery times and can provide the highest certainty for completion on time, experiencing the least delays when compared to traditional contracting and Construction Management methods[300].

In terms of numbers there appears to be evidence from a number of studies both from the USA and UK that actual construction times of Design and Build projects are on

---

[299] See also section 5.2.3 regarding alternative approaches for design completion.
[300] Ling, Khee and Lim, 2001.

average 12 % faster than traditional contracting and 7 % faster than Construction Management methods. Considering the overall delivery period, that is including both the time taken for design and construction, then Design and Build appears to be around 30 % faster than the traditional contracting approach and still approximately 20 % faster than Construction Management systems. Another measure includes the ability for a procurement system to deliver on time, where Design and Build projects are found to be more likely to be completed on time. Results vary from virtually no delay in the US to about 2 % of projects having experienced delay in the UK. The likelihood that Design and Build projects are completed on time appears to be 50 % more certain than with traditional contracting. It is generally the case that certainty of completion on time increases the earlier the contractor is included in the design process[301][302].

What are the reasons for those advantages in the performance of Design and Build in respect of time ? For one, Design and Build is a procurement method that originated with the express purpose of establishing single point responsibility in order to avoid time consuming sequential design, tender and construction[303]. A client's need for urgency may result in the choice of a contractor-led procurement method, since if urgency is a major criterion then it is considered most suitable, particularly if the contractor is experienced with this type of procurement system and is accustomed to the degree of urgency normally attached to such tenders. Also, the difficulties normally associated with subcontractors and project co-ordination and the burden on the client's resources are significantly reduced. So, contractor-led procurement is often considered where projects require multidisciplinary involvement with short construction duration, which is made possible as both design and construction are the responsibility of one entity[304]. This has the advantage that design and construction can be done concurrently, work can be started on site before complete design is available and thus allows for "phased construction" or a "fast-track" approach resulting in a compressed time schedule[305]. Whereas construction management methods can incorporate a fast-track

[301] Construction Industry Institute, 1997.
[302] Ling, Khee and Lim, 2001.
[303] Hughes, Gray and Murdoch, 1997, p. 31.
[304] Smith, 1995, pp. 246.
[305] Halpin and Woodhead, 1998, p. 72.

schedule as well, the risks for early starts on site without completed documents are only assumed by a Design and Build contractor[306].

It is not only the possibility of applying a fast-track approach which enables a Design and Build method to be a more efficient method than other approaches, as the designers, who take most main decisions affecting time, cost and quality, are properly integrated in a Design and Build organisation and thus can operate on a project team basis with those possessing different sets of complementary skills getting to know and respect each other. There are distinct advantages enabling the contractor to use his management skills and experience in the pre-construction period to ensure that design and performance are more closely co-ordinated and better related to time and cost. Economy and efficiency especially should flow from the continuity of joint experience if the contractor is able to offer his services in a series of projects[307]. The effect of shorter times for the design/construction overlap and a design tailored to give the most efficient construction[308] requires a solid working relationship between designer and contractor, whether it is an in-house or external relationship.

Both clients and contractors agree that Design and Build projects can be completed within a shorter time scale compared to traditional projects[309]. A Design and Build contractor commits himself to a completion date at an early stage and the earlier he is included in the design process, on only minimal client's requirements, the greater the likelihood that the project is completed in a short period on time or earlier. However, where the owner's requirements are more detailed, for instance in a case of develop and construct or novation method, fewer projects are completed on time. The more client's requirements are developed, the later contractors are involved in the design and the less opportunity exists to employ the advantageous characteristics of the contractor-led procurement approach just described above.

---

[306] Kubal, Miller and Worth, 2000, p. 336.
[307] CIOB, 1999, p. 100.
[308] Smith, 1995, p. 192.
[309] Kubal, Miller and Worth, 2000, pp. 336; Ling, Khee and Lim, 2001; Seely, 1997, p. 100.

Finally, the Design and Build method is an option available to compress overall design and construction time to accommodate clients as they compete in their own industries. Placing the responsibility for the design and construction with one firm enables the construction industry to effectively include time saving techniques, including fast-track construction, supply chain management, value management and other generic methods, to respond to current time standards. Early involvement of contractors and subcontractors in the planning and design stage contributes to improvements in the overall constructability, quality and increases the opportunity to improve the scheduling process[310]. Completion of projects on time without cost overruns is a feature of BOT projects, since the promoter's / provider's control and continuing economic interest in the design, construction and operation of the project will ensure that revenue flows as soon as possible to minimise costs and risks[311].

4.3.2   Positive features of contractor-led procurement in respect of cost

Perhaps the greatest benefit of contractor-led procurement in respect to cost is that the design should be tailored to give the most efficient construction not only in time but also in costs[312]. This is possible, if it is recognised that contractors, manufacturers and specialist suppliers have a key role to play. They have a wealth of experience, which if brought into play early enough at the design stage, can permit sensible examination of design options and assist in selecting the most cost effective solution to satisfy the client's needs[313]. The lump sum price offered or perhaps a GMP[314] in a Design and Build or Turn-key contract is often considered an advantage by clients who have a limited budget and are not in a position to incur additional costs, as the price is determined at an early stage in the evaluation process[315]. Furthermore, early involvement in the development of a project of an experienced contractor should allow for a less confrontational attitude with the client and his consultants and help in reducing transaction costs through a partnering attitude[316]. The overall reduction in time potentially results in subsequent savings in client's interest costs and he can benefit

---

[310] Kubal, Miller and Worth, 2000, p. 336.
[311] Smith, 1995, p. 264.
[312] CIOB, 1999, p. 118; Smith, 1995, p. 192.
[313] Hill, 2000, p. 7.
[314] See section 3.3.2 for an explanation of Guaranteed Maximum Price.
[315] Smith, 1995, p. 244.
[316] See section 5.1.2 for additional information on "partnering".

from the project sooner due to either earlier revenues or increased efficiency from the completed facility. Although design costs are integrated into the price, they are likely to be less since the contractor-designer compiles essential information only. A firm construction price at an early stage is possible because a single entity controls the design and the construction budget. This reduces the opportunity for variations and thus offers greater security to the client for his financial commitment and the only changes in the scheme for which the client is responsible are those in scope initiated by him – any other are the responsibility of the contractor[317]. Another feature not to be ignored, even if at first glance it appears to be a disadvantage for a client, is the recognition that many firms believe Design and Build projects to be more profitable. If contractors can experience higher profits, clients can expect less cost and time overruns and everyone can benefit[318].

A number of reasons put forward for lower costs arising from following a contractor-led procurement path, include[319]:
- a reduction in construction time (with its attendant cost savings),
- cost effective design incorporating improved buildability,
- use of cost effective materials and construction methods,
- effective use of contractors', subcontractors' and suppliers' resources, and
- the design can be subject to competition, if competitive Design and Build is chosen.

A number of studies have shown that cost savings are indeed possible with contractor-led procurement methods. The Construction Industry Institute has reported that in the USA Design and Build projects had the least cost escalation compared to traditional contracting and Construction Management projects[320]. Another large field study in the USA has been carried out by the Pennsylvania State University College of Engineering, which concluded that Design and Build project unit costs were 4.5 % less than with CM at risk projects and 6 % less than with traditional projects[321]. In the UK a number of

---

[317] Simm, 2000.
[318] Ernzen and Schexnayder, 2000.
[319] Ling, Khee and Lim., 2001.
[320] Erzen and Schexnayder, 2000.
[321] Levey; 1999, p. 239.

studies have reported the same, where the former University of Reading's Design and Build Forum[322] revealed that Design and Build resulted in a 13 % reduction in unit costs and projects were more likely to be completed within a range of 5 % of the agreed budget, with 75 % of Design and Build projects compared to 63 % of traditional projects[323]. Other authors have reported that Design and Build projects are more likely to be delivered to budget[324].

Sources which have confirmed that Design and Build projects are more profitable for contractors include a survey by Practice Management Associates, Boston[325], which reported that 85 % of the contractors questioned stated that Design and Build projects were more profitable, that 15 % thought profits were the same and none thought that Design and Build was less profitable. A more recent study has found results not as favourable, however, in keeping with the normal development of a procurement method into a greater variety of applications which are not all suitable and increasing competition among firms offering Design and Build services, which indicated that 74 % of Design and Build firms believed that Design and Build projects were more profitable, 13 % thought them to be less profitable and 13 % did not have an opinion either way[326].

An in-depth analysis of profitability of Design and Build projects over a period from 1991 to 1997[327] revealed, when comparing profit margins of Design and Build projects versus other projects, that they were 3.5 % higher for Design and Build projects compared to other projects. Civil engineering / heavy type projects led with 9.5 % actual profit margins, building and industrial projects achieved 8.8 % and 6.4 % actual profit margins respectively. Higher profits were due to: better control of the project; teamwork, including people knowledgeable in construction; less competition; negotiated rather than low bid contracts; higher fees to compensate for higher risks; greater design and construction productive efficiency.

---

[322] Now the Design Build Foundation (www.dbf-web.co.uk).

[323] Ibid., 239; Ling, Khee and Lim, 2001.

[324] e.g. Dielschneider, 2000; Halpin and Woodhead, 1998; Kubal, Miller and Worth, 2000; Levey, 1999; Ling, Khee and Lim, 2000; Wong, Holt and Cooper., 2000.

[325] PSMJ Design-Build statistics survey ,1995.

[326] Zweig, White and Associates (Natich, Mass.): Design-Build survey of architecture, engineering, environmental consulting, construction and design-build firms, 1997.

[327] Ernzen and Schexnayder, 2000.

The choice of procurement method may have other more subtle effects upon the efficient running of a project. Where the overall responsibility for design and construction is in separate hands, as it will be in all but contractor-led contracts, the specialist contractor is effectively required to serve two masters. If the enforcement of the design management obligation is left to a party who carries an inadequate level of design management responsibility, the result is easily a network of contractual responsibilities which do not reflect the practical realities of the project. That would be something that detracts from the efficiency of the whole management process.

The issue of whether contractor-led procurement's performance is superior in respect to costs is further complicated when considering BOT projects, which not only include design and construction but also maintenance and operation over a long period of time. Foremost, it has to be said that the provider's control and continuing economic interest in the design, construction and operation of a project, i.e. its whole life-cycle performance, will as a rule produce significant cost efficiencies for the client[328].

Positive features of producer-led procurement in respect of BOT projects include:
- Completion of projects on time without cost overruns to the client (since contractors income is entirely dependent on the performance of the project as early in time as possible).
- Good management and efficient operation (to maximise profit).
- The involvement of the private sector (in public sector projects) and the presence of market forces in BOT schemes ensures that only projects of financial value are considered.
- In overseas work a BOT project can benefit from export financing and can act as a means of financing a project. A firm or capped price can be instrumental in obtaining finance for a project.

There are a number of reports and studies that have attempted to determine the "value for money" of BOT projects in the public sector in comparison to traditional public sector funded projects, which has proved a difficult task since it has not been possible to

---

[328] e.g. Jacob and Kochendörfer, 2000.

have two or more identical projects sourced in parallel under different procurement methods and account for every cost incurred. However, results so far, particularly in the UK from sources such as the National Audit Office[329], the Treasury Taskforce[330] and other organisations, have reported that benefits from a whole life-cycle approach are achieved, especially for those sectors where either large enough scope for improvements in capital investments (i.e. large civil engineering projects) or sufficient scope for improvements in operations (i.e. service intensive facilities such as prisons) exist.

4.3.3   Positive features of contractor-led procurement in respect of quality

Integration of client and producer should reduce the risk of a confrontational attitude and have a positive effect on the overall performance of a contract, including quality standards. The relationship between client, his consultants and contractor can also improve as they work closely on Design and Build projects and all parties gaining confidence in each other. The client is offered a single source of responsibility for the project, which shortens lines of communication at all stages of design and construction and allows parallel working of design and construction (fast-track) which improves buildability and hence quality[331].

"The development of integrated supply chains and construction processes is potentially the means by which the industry will prosper in the 21st century"[332]. One of the principal mechanisms identified to achieve this goal is the early involvement of the construction supply chain in the design and construction process. Integrating the individuals and organisations who can demonstrate the necessary commitment and ability to meet the project objectives improves the flow of information between all parties and to be prompt and accurate. The early involvement of the supply chain in the

---

[329] National Audit Office, 29/11/2001 – "The first ever major survey of central government PFI projects in progress, which reported that most (81 %) public sector bodies involved in PFI projects believe that they are achieving satisfactory or better value for money from their PFI contracts. Over 70 % of authorities and contractors view their relationship as being good or very good with only 4 % of contractors feeling their relationship with authorities was poor."

[330] Arthur Andersen and Enterprise LSE, 2001.

[331] Halpin and Woodhead, 1998, p. 72; Kubal, Miller and Worth, 2000, p. 335; Levey, 1999, p. 238; Simm, 2000.

[332] Hill, 2000, p. 1.

design and construction process, and which contractor-led procurement methods accomplish better than any other, will deliver measurable benefits for the client in improved functionality, improved quality, predictable through-life commitments and meets or even exceeds the client's expectations[333].

Incentives to avoid disputes and to develop innovative solutions to site problems are inherent in the Design and Build type of contract. The adversarial relationship typical of designers and contractors is largely eliminated, since lack of co-operation among members of the design and construct team potentially leads to significant losses[334]. The cause of defects cannot be a matter of dispute[335] and for the reason of a single source of responsibility there is an easy identification of responsibility for any failures, proving to be an incentive for a Design and Build contractor to produce good design and quality workmanship[336]. Of course, such straightforwardness of liabilities in the Design and Build contract in contrast with other procurement systems may mean that strict product liability (i.e. fitness for purpose) could be attached to a Design and Build contract, unless liability is expressly restricted to skill and care only[337]. Experience has shown that the number and type of variations have substantially reduced as well as disputes and claims that often arise under traditional procurement methods[338]. Integrated design and construction may in turn lead to repeat work from existing clients or future work from advisors on behalf of other clients, who are foreign to the Design and Build contractor, by way of referrals, if the quality is right, thus increasing the marketability in the industry. No reputable firms would put their reputations at risk by slighting a customer over quality or related cost issues. Any Design and Build contractor regardless of how the design is implemented, either in-house or externally via alliancing / consortia or subcontracting[339], should produce the highest quality product as contracted. Lowering standards will only result in loss of reputation and future contracts, as it would in any industry[340].

---

[333] Ibid. p. 3.
[334] Halpin and Woodhead, 1998, p. 72; Kubal, Miller and Worth, 2000, p. 335; Levey, 1999, p. 238.
[335] Smith, 1995, p. 192.
[336] Simm, 2000.
[337] Hughes, Gray and Murdoch, 1997, p. 31.
[338] Levey, 1999. p. 238.
[339] See section 5.2.3 for alternative approaches of design completion.
[340] Kubal, Miller and Worth, 2000, p. 344.

The Construction Industry Institute found that there are relatively small differences in the quality of projects procured under Design-Build, Design-Bid-Build or Construction Management. Other US authors also reported that the quality of Design and Build projects is equal to or better than traditional projects. A UK author showed that Design and Build performs consistently better in meeting quality standards in complex or innovative buildings rather than simple and standard traditional buildings. Others suggest that Design and Build produces no worse quality than the traditional system and that there is no apparent reason for quality of construction in Design and Build projects to be lower[341]. It was revealed for contractors to agree that the quality of Design and Build projects is higher than with traditional projects, however, added that this was foreseen as they are not expected to condemn their own quality of project delivery.

There are further benefits from having a procurement process with a single source of responsibility in that Design and Build ensures the client for project drawing files, usually provided today in an electronic format (CAD)[342], to be available for use throughout the entire design and construction process. Something that has yet to become a standard feature with traditional contracting methods. Design and Build facilitates the computerisation of the construction process, which demands that CAD drawings and files become the nucleus of project communication and data files, including the linking of estimates and schedules with CAD files, to improve the overall quality and timeliness of project completion. Design and Build creates a contracting method that can utilise network linkage of CAD files as a common denomination to improve the overall construction process, since a single source is responsible for the whole of the design and construction process, only accountable to the client in respect of a particular project[343].

Another positive feature of producer-led procurement to mention are standard building systems developed by producers / contractors, where the use of the Design and Build procedure can be beneficial. If a producer's / contractor's proprietary system can be used without detriment to the client's requirements, there can be economic and quality advantages in the use of a Design and Build method, preferably incorporating choice of

---

[341] Ling, Khee and Lim, 2001.
[342] CAD = Computer Aided Design.
[343] Kubal, Miller and Worth, 2000, p. 337.

layout, finishings and external works[344]. Benefits from standardisation, pre-assembly and modularisation, where adoption is facilitated by producer-led procurement methods, in respect to quality are better risk control and reliability, safe working practices and less on-site problem solving, increased reliability of building performance and higher quality of work in both aesthetics and appeal[345]. Reasons are off-site improvements in manufacturing, quality of modular components, reduced repairs and maintenance costs of modular buildings, achieving consistent levels of quality with less snagging and manufacturers guarantees on pre-fabricated modules much longer than normally expected for a building[346].

The issue of quality when considering Turn-key or BOT projects is of particular importance, where construction of a turnkey contract will be carried out by the contractor who will in most cases operate it for a period of up to two years after commissioning, the client then taking over the operation for the life of the facility. In BOT contracts the contractor will usually enter into a concession contract for a much longer period before finally transferring the facility to the client. The producer's design obligation, irrespective of the specification adopted, will form part of his general objective to supply a facility that meets the required performance specification and guarantees to the client. Especially in respect of BOT contracts, where the producer owns and operates the facility for the duration of the concession on behalf of the client and collects revenues in order to repay the financing and investment costs, maintains and operates the facility and wants to make a margin of profit, he will ensure an appropriate level of quality of both design and workmanship that satisfies all these demands[347].

---

[344] Seely, 1997, p. 99.
[345] Cox and Townsend, 1998, pp. 258.
[346] David Langdon & Everest, 2002; Gibb and Isack, 2001; Hughes, Gray and Murdoch, 1997, p. 59.
[347] Smith, 1995, pp. 243.

4.4   Less favourable circumstances of contractor-led procurement

4.4.1   Circumstances less favourable for contractor-led procurement in respect of time

Whilst the majority of studies and most authors agree that contractor-led procurement offers a better performance in respect of time than other procurement methods, there are some that suggest that Design and Build is no quicker than a conventional project. Even when design and construction periods in Design and Build are shorter, the scope development stage / concept stage and team solution period are both claimed to be longer than under traditional contracts. Longer time may be needed to draft performance specifications and client's brief carefully. A longer tendering period for contractors is required, where in traditional contracting bidders are given three to six weeks to submit a tender and in Design and Build projects three to four months, sometimes up to nine months, of tendering period is recommended. A longer time period has of course to be allowed for contractors to develop the design concept in order to submit realistic tenders[348].

These criticism viewed in isolation are correct, particularly when simply comparing construction or design stages, but the overall effect of single responsibility in the original modes of contractor-led procurement offers time benefits. The better the brief has been developed at the outset, the more likely that the project will be a success in meeting the client's objectives and the lower the chances that increases in cost and time occur because of changes during the construction phase.

A more serious criticism of contractor-led procurement systems in respect to time is that the objectives of the Design and Build firm may at times conflict with those of the client, where for instance speed versus construction method, speed versus best design solution or speed versus economy pose problems that will have to be resolved, depending on the relationship between client and contractor, his contractual position and his expertise[349]. This is not to say that these are problems not encountered with other procurement systems, only that the client has usually reserved himself a more

---

[348] Ling, Khee and Lim, 2001.
[349] Walker, 1996, p. 212.

flexible position, which to a large degree explains many of the problems associated with traditional or management methods of procurement. In Design and Build the client is in a relatively weak position to negotiate change after contractual close, since he has committed himself to the whole package, including time to completion, at an early stage[350].

The comparison and evaluation of tenders is more difficult due to the possible variation in the design concepts and information submitted by contractors. This can lead to numerous post-tender enquiries if not fully analysed at pre tender stage[351]. At the same time, the client receives a number of different proposals and costs, allowing a number of options to be examined. The options would give the client far more flexibility over the traditional type of contract based on one design. The more detail of clients' requirements are provided, the most extreme situation is encountered with the novation method of develop and construct Design and Build, the fewer projects are completed on time or earlier, simply because the later the contractor is involved in the design under such circumstances, the less opportunity for time saving techniques on the basis of integration there is, and the greater the risk he carries that he has misinterpreted aspects of the design from the intention of the client or his consultants. Competitive Design and Build, for both design and price, is likely to involve the longest time taken for the tendering period[352]. Time will be lost that could have been spent in useful and productive negotiations, which are likely to arise with competitive Design and Build as well, perhaps even more so.

4.4.2 Circumstances less favourable for contractor-led procurement in respect of costs

Several studies concluded that Design and Build projects are neither cheaper or more expensive than traditional projects[353]. Explanations include the opinion that Design and Build projects do not cost more or less than traditional projects because the same amount of work needs to be carried out. It is thought that in practical terms, the

---

[350] Smith, 1995, p. 192.
[351] Simm, 2000.
[352] Walker, 1996, p. 211.
[353] Ling, Khee and Lim, 2001.

financial advantage of Design and Build projects is difficult to quantify, because clients do not call one tender based on Design and Build contractual arrangements and another tender based on conventional methods or other just to compare which arrangement is cheaper. Actually, this has been done in Germany by local authorities on occasion in order to see which approach produces the best value for money approach. In most cases, unfortunately, the comparison made between different bids did not happen on an equivalent basis, where the traditional (separate trade contracting) bids, not surprisingly, came in lowest in cost, since trade contractors had to undertake the least amount of bid preparation, bore a minimum of risk and thus experienced the lowest on-costs compared to general contracting on a lump sum basis, Design and Build method or package deal. A further problem in warranting an objective and unbiased evaluation of project performance is that only tenders at award stage were analysed, neglecting any follow-up project appraisal during construction or after completion, never mind operation. Therefore, any measure of actual performance in respect to time escalation, cost overruns or quality deficiencies was impossible[354].

If clients feel that they are going to make changes to the design during construction, whether the contract is based on Design and Build or other procurement methods, then the contract sum is bound to change and it is not realistic to expect price certainty from a Design and Build contract. There exists little flexibility to accommodate variations with the traditional Design and Build method after the price has been confirmed. This can mean greater cost implications if variations in scope do occur[355].

It may be difficult to identify design elements in tenders which could cause future maintenance problems for reason of limited design information submitted at tender[356]. Savings in design costs that a client expects could be offset by the need to employ additional professional expertise to prepare the tender documents and police the work.

---

[354] Wlassak, 2001.
[355] CIOB, 1999, p. 117.
[356] Something that can be overcome if Design and Build is extended to include operation and maintenance by the contractor under BOT / PFI projects.

The implication of a contractor adapting to the demands of a Design and Build tender bid does mean an increase in head office costs, where the contractor requires additional staffing levels, which may be either in-house or external, dependent on the contractor's resources or business strategy. These include: design and build co-ordinator, designers, specialist engineers and quantity surveyors. The result is a higher overhead percentage level than that of contractors bidding for traditional contracts.

Reasons cited, therefore, to argue for higher costs of contractor-led procurement are[357]:

- higher risks on the part of contractors,
- pricing based on incomplete drawings,
- higher overheads for contractors due to the early and greater involvement,
- higher profit margins,
- additional set of consultants may be employed by clients to supervise contractors and their consultants, and
- additional insurance coverage for the contractor and higher bond rates.

The extent of competition is likely to be reduced in a particular market especially at the outset of contractor-led procurement[358].

In respect of BOT projects it has to be recognised that it involves a highly complicated cost structure, which requires time, money, patience and sophistication to negotiate and bring to fruition. From a contractor's / provider's perspective the risk associated with BOT projects are far greater than those considered under traditional forms of contract and still more than those mentioned under traditional Design and Build projects as the revenues generated by the operational facility must be sufficient to pay for design, construction, operation, maintenance, finance and investment in unsuccessful BOT bids. The uncertainty of demand or level of performance and hence revenues, cost of finance, length of concession period, levels of tolls and tariffs, effects of commercial, political, legal and environmental factors are only some of the risks to be considered by producer led, promoter organisations involved in BOT style projects[359].

---

[357] Ling, Khee and Lim, 2001.
[358] Smith, 1995, p. 192.
[359] Ibid. p. 265.

With shorter tender periods imposed by clients, the contractor's bid team are under increased pressure to deliver the correct judgements that will be successful by winning the project on costs, innovations and forecast profit margin in competition.

4.4.3 Circumstances less favourable for contractor-led procurement in respect of quality

A number of authors and surveys of clients have raised doubts about the quality of Design and Build projects. The most common critique is that many Design and Build firms come from a contractor background, which will give precedence to construction rather than to design quality and as firms have a tendency to be orientated towards construction activity, that it will have detrimental consequences for the integration of design and a subsequent effect upon the quality of a project[360]. It is claimed that the Design and Build firm may not always act in the client's best interest[361], that quality may be compromised in Design and Build projects and that clients express more dissatisfaction with Design and Build projects than with traditional projects on account of poorer quality or receiving only the lowest acceptable quality[362] and the inability to meet functional requirements[363]. Two factors only are thought to affect design development in Design and Build projects: one is to meet the client's requirements, the other to design so as to reduce the contractor's cost. It is the second factor that is often thought to cause financial pressure on the contractor, which leads to a reduction in quality, since it is the single minded aim of the contractor to reduce cost by providing cheap solutions for the achievement of higher profits. It has been suggested that it is difficult for contractors to represent clients' interest and that their only interest at the same time becomes profitability taking precedence over design. This represents a mind set that views each project as if it existed independently and isolated of anything else, that a contractor was free to maximise profit to the detriment of any other aspect and as if there was no competitive market or indeed any future at all, which a contractor has to consider at all times. It is true that a rogue or perhaps inexperienced contractor, not realising the implication of Design and Build, may take such an attitude and manages to

---

[360] Ling, Khee and Lim, 2001; Simm, 2000; Walker, 1996, p. 212.

[361] Seely, 1997, pp. 97.

[362] Ling, Khee and Lim, 2001.

[363] Dielschneider, 2000, p. 29.

convince an inexperienced client to accept its services, only to find itself most likely out of business very soon after. A particular danger exists if an inexperienced contractor offers a Design and Build service, who does not have a track record of either Design and Build or of the type of building in question, and does not realise the effort and input required or risk committed to, especially in terms of management and communication skills required between all contributors. Thus, he fails to satisfy the client[364].

Another area often mentioned to be a weakness of Design and Build procurement is that it offers little flexibility if requirements are changed[365], where the client is in a relatively weak position to negotiate change[366], variations may proving to be expensive and the quality of the work likely to suffer[367].

Many of the problems referred to above stem from a poorly developed or ambiguous brief that is to define the task of the Design and Build contractor and provides the basis upon which he locks into the contract with price and time commitment. Many unsuccessful Design and Build or turnkey contracts have resulted from an inadequate definition of requirements at tender stage. A poorly defined brief makes it extremely difficult, if not impossible, to compare accurately the bids received[368]. A definitive statement is not to mean a fully developed brief in terms of design and a detailed specification including measurements, but is to be considered as a clear statement of objectives and criteria to be met for the satisfactory operation and fulfilment of its fitness for purpose. It is, therefore, necessary to have the client's requirements documented in a definitive statement early on in the process as it becomes the basis for all subsequent activities[369]. Even then, the evaluation of contractors' tenders can be complicated as each contractor is likely to interpret the brief in a different way or the evaluation can be very subjective when aesthetic aspects are important[370]. Little or no influence in selection is possible in Design and Build for selecting preferred trade or

---

[364] Dielschneider, 2000, p. 30.
[365] Ibid. 29.
[366] Smith, 1995, p. 192.
[367] Seely, 1997, p. 98.
[368] Smith, 1995, p. 243.
[369] CIOB, 1999, p. 47.
[370] Seely, 1997, p. 98.

works contractors which are to actually execute the works[371] after the contract has been signed.

Designers have thus been reluctant to advise the use of Design and Build methods for projects where the design is of paramount importance to the client, where he probably wishes to choose the architect independently or by means of an architectural competition and would not want to be tied to a single contractor. Refurbishment work rarely lends itself to this type of arrangement, and clients requiring purpose made buildings will generally prefer an independent design team and select other procurement options that allow progressive development of the client's brief, proving to be helpful where there is uncertainty or greater complexity involved[372]. There are a variety of ways of involving contractors at an early stage to work with the design team[373].

Other details cited which militate against Design and Build procurement methods are that some clients feel contractors, who are profit driven and need to be responsible for a building only during the defects liability period, not to bother to take maintenance issues into consideration[374]. It may be difficult to identify design elements in tenders which could cause future maintenance problems due to the limited design information submitted at tender. Both these issues are overcome in Turn-key projects with some operational requirements and BOT projects. Another feature that separates Design and Build from other construction procurement forms is the lack of an independent certifying role for the lead designer[375]. There usually is no provision for independent monitoring of construction quality and if any monitoring is deemed necessary by the client it must be independently commissioned[376]. As a consequence the client is placed in the position of a "policeman", whereas under conventional contracting the architect acts as the client's agent and controls the contractor to a certain degree[377]. It is argued that checks and balances present during the conventional method are not present in

---

[371] CIOB, 1999, p. 117.
[372] Ibid. p. 117, Seely, 1997, p. 98.
[373] see chapter three.
[374] Ling, Khee and Lim, 2001.
[375] Hughes, Gray and Murdoch, 1997, p. 31.
[376] CIOB, 1999, p. 117.
[377] Levey, 1999, p. 239.

Design and Build processes[378], causing the final product to be below the client's expectations.

While there is generally agreement between clients who use Design and Build to benefit in terms of time and cost, there is disagreement on the level of design and build quality which they will receive[379]. Often, however, problems arise from differences between the contract conditions and the client's requirements, because the brief and tender conditions were not clearly defined[380]. Therefore, Design and Build methods have usually been advanced for projects where the design is uncomplicated and innovative solutions and processes would be inappropriate[381]. This advice is, however, not in keeping with the origin of Design and Build for multidisciplinary and complex industrial facilities, nor does it properly take into account the integrative powers of the Design and Build process.

4.5   Appropriate application of contractor-led procurement

It is readily apparent from the investigation of positive and less favourable features of contractor-led procurement that those characteristics which support the selection of a Design and Build, turnkey or BOT approach are also at the root of the perceived problems associated with it. If the client expects cost and time certainty from a contract with risk of project success transferred largely or almost entirely to the contractor / producer, then he looses a degree of control over the project in terms of design development, variations during design and construction, influence over the processes and actions adopted by the contractor. The rules of the project must therefore have been laid out in the brief and agreed upon at contract signature in all respects as any changes to the rules thereafter causes conflict, which can be detrimental in both cost and time and possibly could endanger project success. This implies that the client knows what he needs at an early stage and is capable of communicating those needs to the contractor / producer avoiding ambiguity and leaving no room for misinterpretation. This usually calls for a reasonably experienced client, or a contractor that can be relied upon to act in

---

[378] Ibid. p. 289.
[379] Ling, Khee and Lim, 2001.
[380] Smith, 1995, p. 249.
[381] Hughes, Gray and Murdoch, 1997, p. 31.

the best interest of the client. There are a number of alternatives at which stage and on what basis a contractor / producer can be brought in, giving rise to a variety of producer-led procurement methods described already in section 3.3.5 with their attendant advantages and disadvantages explained hereafter.

4.5.1 Analysis of positive and less favourable features of contractor-led procurement

The previous discussion showed, that there is generally agreement on contractor-led procurement, either in the shape of Design and Build, turnkey or BOT, to deliver project results that are usually time efficient, frequently offering cost economies and allow construction expertise to be integrated with design at an early stage. However, contractor-led procurement is thought not necessarily to provide the same degree of design or build quality as other procurement routes, especially management-led methods. It is also recognised that a conflict exists between client's needs and what they expect from a procurement delivery method, where for example wholesale transfer of risk to the producer limits the influence of design control or flexibility after the contract has been signed. A management-led approach, which allows for controllable flexibility, has the disadvantage of risk remaining with the client

It is those features that have led to a variety of contractor-led procurement forms, essentially representing a sliding scale of involvement of the producer / contractor from very early to fairly late in the design process and the degree of involvement after completion of the project.

Beginning at one end of the spectrum, it is the BOT approach that delegates the maximum of responsibility for a project over its life-cycle from as early on as possible. It is also the most complex of situations, where the brief has to involve all aspects of the project over its whole life-cycle, in cost, quality and legal terms. It requires the client to be particularly experienced and knowing what it is he requires from the project, thus depending on expert advice from a variety of consultants including legal, financial, technical and design advice. In this type of arrangement there is a strong incentive for the producer / provider to deliver and maintain a facility at an acceptable level of quality, in order to protect his revenue stream as previously explained in section 3.3.5. To optimise overall project quality, especially economic efficiency, it is imperative for the producer to be included immediately after the briefing stage, so as to be in a position of control over design development, as it directly affects his position in the subsequent construction and operation phases. The better he has managed to achieve overall project quality in all respects, the better his economic position as determined by the client in terms of the concession. It involves the highest costs for bid preparation of any procurement route as the greatest effort is demanded to encompass all aspects over the full life-cycle of a building in a sophisticated bidding process.

A Turn-key package is rather similar to a BOT project, but with one major difference in that the producer / contractor usually hands over the building / facility after commissioning has been completed. The risks associated with ongoing operation thus remain with the client, thus it is of utmost importance that the contract conditions reflect accurately the client's requirements and the Turn-key contractor carrying out the works has a good track record and provides worthwhile guarantees.

The method of direct or traditional Design and Build offers to the client the best potential of benefiting from the advantages discussed earlier. Early involvement of an experienced contractor with a proven history in the type of building required allows for a maximum of improvement of processes involved in designing and completing construction projects by growing a team that can more effectively share resources, capabilities and ideas to improve costs, time and quality. It also requires for the client to trust the contractor that he will act in his best interest and in return manages the design and construction risks on his behalf. It is this method that requires the highest degree of trust on the part of the client that the design solution and the price do indeed reflect that is possible to satisfy his needs. Again, it requires an experienced client to ensure that the service and product offered by the Design and Build contractor meets his requirements. One of the ways to ensure that a client has access to this type of experience is the employment / appointment of a professional project manager, who specialises in the type of project in question. It is with the introduction of a professional project manager that the client has someone competent to look after his interest, to ensure that the

Design and Build service delivers upon its promises. Such a consultant can be appointed to act on behalf of the client and to advice the client with the arrangement for tenders for the work to be submitted and then for the evaluation and selection of a suitable contractor to fit within the client's expectations. An independent advisor[382] can also monitor quality and cost, not only during the tender and design stage, but also during the construction phase, thus overseeing the whole process[383].

While the Design and Build process potentially provides the most effective integration, there remains the difficulty of effective integration between project team and the client. The client needs to protect its position so that the project it receives on completion fulfils its requirements. The client must be in a position to resolve situations to its benefit if it has sufficient in-house expertise to understand the issues and the appropriate contractual conditions that allow the client to act to produce a result to its benefit. If the client has not, then professional advice upon which to act will be needed. Professional advisors in this capacity would act as a substitute for the client's in-house project management team[384].

Whereas direct or negotiated Design and Build allows for a team approach, provides some flexibility on the part of the Design and Build activities during the development and construction process and created the kind of circumstances most suitable for an integrated approach, it requires at the same time from the client the fullest of trust and belief in the contractor in respect of cost, time and quality. Other, not as experienced clients, or perhaps not as sure of what is required to meet their needs, or need to demonstrate lowest price such as the public sector, decide on a competition of design and price. It is costly for Design and Build contractors to tender in competition as each contractor will have to produce a design to meet the brief and a price for construction. Of course, these will demonstrate whether a Design and Build contractor is sincere about his tender and it allows the client to choose among the most suitable propositions. Where, however, this process is taken to excess at the tendering stage, it will result in an

---

[382] See also comments regarding the nature of an independent advisor in section 3.4.1.
[383] Cox and Townsend, 1998, p. 37; Dielschneider, 2000, p. 30; Ling, Khee and Lim, 2001; Pilcher, 1997, p. 28; Seely, 1997, p. 99.
[384] Walker, 1996, p. 214.

unnecessary use of resources, if for instance more than three or four firms are expected to go the whole length of the tendering process. Hence most contractors are not prepared to go beyond outline sketch design and an indicative price[385].

To obtain the full potential of a Design and Build procurement system, it is preferable to have only a written scope package prepared to permit proposing Design and Build firms to submit proposed conceptual designs. This design is often the first phase of a market-place selection process, where the client will review the first phase of a multiple-phase selection process. After review, the client should shortlist firms that provide acceptable conceptual designs and only invite those selected, ideally three to four firms, to submit further more detailed designed and cost proposals. Some clients abuse the interest of Design and Build proposals by requiring extensive design submittal packages from all proposers with no regard to the cost borne by them. Some clients will not attempt to limit or shortlist the competition to those that have a reasonable chance of being successful[386]. Contractors thus exercise diligence in choosing appropriate Design and Build opportunities for submitting proposals and most contractors formulate strategies to determine the optimal number of proposal submittals based upon past success ratios and corporate budget allocated for marketing. Due to the high cost of Design and Build proposals, most contractors will elect not to submit on potential projects unless they have more than a reasonable expectation of preparing a winning proposal. This expectation ratio ranges from a low of 33 % to a high of 50 % or above with anything lower typically rejected depending on the contractor's current and expected workload[387].

One of the perceived disadvantages of Design and Build is that clients are afraid to loose control over the design or perhaps it is the clients' traditional consultant, the architect, who is afraid of loosing influence. Many industry customers want to benefit from the transfer of risk to the contractor but at the same time maintain control over design, effectively recognising Design and Build more as a means to transfer risk contractually rather than as a method of improving the construction process. This is

---

[385] Seely, 1997, p. 99.
[386] Kubal, Miller and Worth, 2000, p. 354.
[387] Ibid. p. 355.

accomplished by the client having an architect for completing a portion of the project's design before awarding the Design and Build contract. It is common practice for clients to appoint an independent designer to prepare schematics and design parameters on which Design and Build proposals are to be based[388]. Some clients, however, take this initial design stage to the extreme, in some cases actually completing the entire architectural design and requiring the "Design" and Build contractor to prepare only the associated engineering designs, such as structural and services design, and assume all risk for the design work already completed. In this situation the contract cannot be considered Design and Build but rather a Design-Bid-Build with assumption of design risks[389]. Naturally, clients often wish to maintain a large share of design input, but by completing the design to the extent that the contractor has no ability to provide process improvements defeats the advantage of Design and Build contracting.

This approach taken a step further is represented by the "novation" Design and Build method where the client's designers, who have developed the project to the point of appointment of the Design and Build contractor, are passed to the contractor for the completion of the project[390]. Now the contractor is not only responsible for the design so far, but is additionally accountable for the performance of the designer for the remainder of the project and has to face all the organisational difficulties that such an approach can pose[391].

4.5.2 Two examples of Design and Build projects

Office development in Salford, United Kingdom[392]

The project was to develop serviced new office accommodation of approximately 22,000 m² of floor space on seven floors for 1940 staff workstations, including basement car parking for 180 vehicles, a restaurant, gym and conference room, for the Inland Revenue in Salford under the Private Finance Initiative. The client / principal being the Inland Revenue awarded the concession to the promoter / provider represented

---

[388] This process is also known as "develop and construct".
[389] Kubal, Miller and Worth, 2000, p. 353.
[390] Barnes, 2001; Sterling 2001.
[391] Walker, 1996, p. 211.
[392] Masters, 1998, pp. 10.

by London & Regional Properties, who in turn let the design and construction contract to Balfour Beatty Construction.

The invitation to tender for the contract was issued on the 3rd of June 1996 and a design proposal for the project was submitted by Balfour Beatty on behalf of its client on 17th of June 1996, after an intensive 14 days of activity including planners, designers, architects and construction personnel. Following the award of preferred bidder status to London & Regional Properties on 17th of April 1996 the design development and planning process were carried out in earnest with the contract signed one week before Balfour Beatty moved onto site on 14th of April 1997. It was this period of design development which showed the strength of the Design and Build approach, where consultations between the client and Salford City Council's planning department resulted in the required site and with it the layout of the building to change.

The initial design of a rectangular building facing the water front was to straddle the extended axis of a show-piece cable-stayed foot bridge designed by Spanish engineer Santiago Calatrava. This subsequently became unacceptable and the permitted rectangular site boundary was rotated and moved to the side of the extended axis. As a consequence of this the design was changed by Balfour Beatty so that the building is now a modified "H" shape with one leg shortened and arranged at a skew from the rest of the building to run parallel with the bridge axis. The change of design was accomplished without affecting the contract completion date and budget, with a handover date of 17th of August 1998, being that of service availability, thus including the entire fit-out and commissioning of building services and a fixed budget of £ 25 million for the capital cost of the construction work. At the same time as keeping one eye on these activities, the contractor's designers and planners have ensured that the building will fit the end user's requirements.

Other changes that took place on account of involving all parties affected by decisions made during the design process, which revealed that due to the nature of the work a fair degree of privacy is required, was to increase the cellularisation of an open plan office originally envisaged. Consultation with seven different user groups within the Inland

Revenue and union representatives enabled them to have an impact into fittings chosen, which resulted in proposals for full vending facilities to be scrapped in favour of kitchen facilities being incorporated on each floor instead. This was all part of Balfour Beatty's contract, including in fact the entire fit-out of the building. The period between practical completion and service availability still exists, but within the contractor's contract completion date, so that when the end user moves in all fittings, furniture and computer equipment will be in place.

These changes and a construction method optimised as to meet the tight time schedule, where a self supporting shell frame was chosen by the design contractor to preclude the need for stability from lift shafts, allowing 2000 tonnes of structural steelwork to be erected in just 16 weeks and enabling the building's plant to be placed at the top of the building soon after. Thus minimising crane usage and moving external cladding off the critical path of the construction programme by opting for a self supporting frame and lightweight metal studding with aluminium panels to enclose the building quickly from the top, and allowing work to continue simultaneously on several fronts and have internal services work begin early. All of this is made possible by the Design and Build method of construction procurement as it enabled Balfour Beatty to capitalise on its own culture of close liaison and co-operation between all parties and manage a near impossible task of changing a site while keeping to schedule and within budget.

Balfour Beatty had acted on the bad reputation that Design and Build had gained among some clients, because construction teams were failing to be responsive and flexible to change or were not including the end user of the project sufficiently during the design and construction process. Clients typically used to comment that they were not fully aware of what they were getting until the finished product was presented to them. Balfour Beatty ensured that the design brief was understood by all, validated its design externally and presented it to the client after carrying out an internal audit. Effective control of the design process was made possible by engineering the process to accommodate changes which are inherent and usually abundant in any construction project. Good IT systems have been invaluable for managing change in terms of updating documents and drawings efficiently and applying cost control to accommodate change. IT also allows the longer term implications of change to be evaluated at the time it is made and show clearly to the client that different options are being considered. Design and construction proved to be a near seamless process with much less departmentalisation to enable everyone to share the same objectives. Most importantly, people that would previously only have been involved in the construction process were brought in at a much earlier stage and architects were given the role of design management from pre-tender right up to the point of handover, thus crossing the divide that previously existed at he point of contract award.

Brindley Place development in Birmingham, United Kingdom[393]
Here the Design and Build contractor HBG has completed a number of significant buildings on the Brindley Place development in Birmingham, working with the client Argent group PLC. These include Three and Five Brindley Place and Bar Rouge, an acclaimed glass-clad restaurant by Piers Gough.

Five Brindley Place was already pre-let to British Telecom and the first detailed cost plan by HBG when negotiating with the client in 1994 showed that the project cost had risen £ 2 million above the £ 14 million budget approved at pre-let stage. However, the establishment by HBG of a multi-disciplinary in-house team dedicated to undertake client's Argent projects supplemented with staff from key specialist subcontractors[394], recruited on the strength of their experience and capacity for innovation, proved the key to success. Every member of the team was encouraged to contribute, discuss and modify plans. A conflict-free contracting environment resulted, where problems, which inevitably occur with any construction project, were resolved before they impacted on progress on the basis of close working relationships. So, a series of design changes to the project by the Design and Build contractor brought the cost down to pre-let level and construction started and finished exactly on programme with the final account agreed on the last day. Again, incorporation of variations as a result of discussions with the client and end user was facilitated by the use of IT to its full potential, where HBG modelled everything on computer and analysing each output of the design before

---

[393] Howell, 1999, pp. 34.
[394] See also chapter six for more information.

moving to the construction phase. A major benefit to both client and contractor as a result of their ongoing relationship of six years with seven buildings completed and more under construction with a total contract price of £ 80 million, is that earlyinvolvement in the design process has enabled the team to identify, research and develop efficiency gains and apply continuous improvements, which has resulted in a 10 % reduction in cost and time for a typical 6,500 m² office building. Techniques adopted include value engineering, research and investigating the use of prefabricated items and the deployment of the multidisciplinary team on all projects[395].

The key to success in those projects, as already explained earlier, is the scope that Design and Build offers for involving everybody concerned with the project as early as possible in the development process. "The most innovative clients recognise the benefit of including as many of the team as possible, right from the beginning, and innovative contractors realise the importance of design". However, the common Design and Build practice of novation, where the design team is employed first by the client to prepare the scheme up to funding stage and then passed over to the contractor to work on the construction of the project is not the best way to proceed. "Designer and contractor are still separated, so where's the advantage?"

4.5.3 Preferred application of contractor-led procurement

Having considered the benefits and constraints of contractor-led procurement as well as procurement issues from a client's and project's perspective it is necessary to consider its preferred application so as to enhance its performance and choose a variant of producer-led procurement that satisfies as many of the client's priorities as possible. Applying the headings of client's priorities established in chapter 3.4 to the strengths and weaknesses of producer-led procurement it is now possible to consider the implications of all the variants and recommend the best course of action.

Price certainty

There is general agreement that contractor-led procurement offers the greatest degree of

---

[395] See also sections 5.1.2 regarding experienced clients and section 6.3.4 regarding early involvement tools.

price certainty among all of the alternative procurement options. Price certainty can be achieved relatively early in the project development process, depending in some measure on the level of trust between client and contractor that each is representing the full picture from the outset and that neither the client, in demanding by way of his prerogative to authorise design a higher than anticipated standard, nor the contractor, by failing to deliver on his representations, jeopardise the trust placed in each other. Clearly, a good reputation or an enduring relationship is very supportive in this matter. It is in the best interest of the contractor / provider; Especially under BOT arrangements; to optimise his costs over the life-cycle of a facility and in return receive as much profit as is possible under competition and constraints placed upon him by the contract / concession.

Timing

If there is one aspect of contractor-led procurement of little doubt it is that of timing, where it is widely accepted that timing either in terms of certainty for completion on time or completion in as short a time as possible is a factor strongly in favour of contractor-led procurement. This is not to say that other procurement options, especially management –led methods, cannot deliver in this respect, but contractor-led systems display a performance at least equally as good. However, there are differences in the ability of contractor led procurement methods to be a time efficient method of delivering to construction needs. The greater the design development in a prescriptive manner by consultants on behalf of the client, the less the opportunity for the Design and Build contractor to optimise the remaining design and subsequent construction. Whilst it is important for the client to be clear on his needs and have these communicated effectively and in a precise manner, it must be performance related as far as practicable and not prescriptive by way of drawings, details and bills of quantities. On the other hand it is for the Design and Build contractor to demonstrate a sufficient degree of flexibility in his ability to translate these requirements into a satisfactory design solution and up to him to ensure close client participation including critical members of the supply chain. It is readily apparent that timing and controllable variation, the next criterion, are closely interrelated.

Controllable variation

The analysis and case studies have shown that experienced contractors utilising computer technology for the design and planning process are relatively versatile in adapting a design to meet client's requirements, as long as its basic project parameters remain true to the initial brief. However, substantial variations to either scope (i.e. size, quality, purpose) and timing of execution of the project will most certainly affect the cost and time guarantees given by the Design and Build contractor earlier in the development process. If such uncertainties exist at the outset, prior to embarking on a project, then it is unrealistic to expect any certainty from the project and a reactive style of construction procurement becomes necessary, as offered by management methods which are infinitely flexible, however, do not give cost and time certainty. Direct / negotiated Design and Build thus offers the greatest degree of flexibility within any of the contractor-led procurement types and in a situation where a client is not as experienced or as sure as to what precisely he requires and needs advice with the arrangement of the brief, for tenders for the work to be submitted and selection of a suitable contractor, then he should employ the services of an experienced professional project manager in the type of project planned.

Complexity

Although there is some criticism for the case of Design and Build to be thought of as appropriate for complex projects, this is not reflected in Design and Build's origin of multidisciplinary and complex industrial projects, which have benefited from an approach of single source responsibility for project development, construction and delivery, particularly in a turnkey format which includes commissioning. Even more risk is transferred with BOT style construction procurement for project delivery, where payment is only due in return for specified services or output. Particularly, where proprietary technology of either process or construction knowledge, or simply a good track record for delivery of a certain project type is desired, a contractor-led approach is beneficial. Once again, it is either a direct / negotiated approach which best serves the needs of the client for transfer of risk to such a degree that payment will only flow on receipt of a serviceable building or facility. Intermediate forms run the risk of a clash of interest and responsibility occurring, especially under circumstances of greater complexity.

Quality level

Whereas workmanship, functionality or fitness-for-purpose can successfully be provided for in contractor-led procurement, if the brief has been satisfactory in precisely expressing a client's requirements, it is quality of design which is frequently under debate in contractor-led procurement systems. Particularly for those types where the contractor bears the greatest amount of responsibility, i.e. direct Design and Build, Turn-key and BOT, he is most often criticised for poor or indifferent design. Surely, guarantee of successful design can never be granted under whatever procurement path is chosen. Nevertheless, the opinion of most designers tends to support the view that an independent architect, directly contracted to the client, provides the most suitable framework to achieve an aesthetically and architecturally successful project. Whether the same holds true for of achieving functionality or fitness-for-purpose in a project cannot be so readily confirmed.

Where the architectural design is of significant importance to a client the best compromise in terms of a contractor-led approach is to expose the contractor selection process to a design competition. The submission of proposed conceptual designs represents the first phase of a market place selection process and enables the client to shortlist only those firms that on the one hand offer suitable credentials and on the other hand an interesting design solution, perhaps in co-operation with a known architect. This approach should not be abused by a client, as it will only result in putting-off potential bidders if they are expected among many others to prepare fully detailed proposals right to the end of the procurement process only then to learn that they have not been successful and all their effort wasted.

Simply transferring the risk for design to a Design and Build contractor after having substantially completed a design with designers directly appointed by the client in a process of design novation may appear to offer the client the best of both worlds, however, frequently fails to deliver upon its promises for a number of reasons, chiefly because this process factually persists with the separation of design and construction responsibility, whatever the wording in the contract. As an example, a lawyer commenting on this type of arrangement has said, that "employers, of course, want to have their cake and eat it and impose on contractors, in this hybrid situation, the full risk of a proper Design and Build contract. Contractors should resist this. Otherwise, they will need to make a decision as to whether they have to (rework the design), or whether they price for the risk"[396].

If the client is certain that only an independent architect can fulfil his architectural needs, then contractor-led procurement is not the ideal form of construction procurement and a management-led approach is a better choice. The client, however, should be aware that in this case he cannot expect price and time guarantees at the outset and is better off ensuring to have sufficient time and funds available.

Contractor input

Obviously, Design and Build, especially of the direct type, represents the organisational format best suited for contractor input; more direct than any other procurement system. It is this feature, especially integrated design and early involvement of strategic suppliers, that explains to a large degree the benefits to be had by choosing a contractor-led approach to construction procurement and offers a client cost and time guarantees at an early stage in the project development process. Should the client not be as experienced or as sure of how to communicate or relate to a Design and Build contractor, he is best advised to include a professional project manager to act on his behalf and lend his expertise to ensure a successful project outcome. The later the contractor becomes involved, the less opportunity he has to optimise the design and construction process and to ensure a good value for money return.

It is up to the contractor / provider in BOT projects to ensure that overall life-cycle performance is optimised so as to achieve a maximum return on his investment over a longer period of time in fulfilling the requirements of his contract / concession.

---

[396] Barnes, 2001.

Competition

Contractor-led procurement in its direct form of negotiated price offers to the client the greatest degree of price certainty at an early stage in the project development process. This is a suitable approach for the client who is aware of what the market demands in return for the services and product required, but not as suitable for the client who is not as experienced or must outwardly demonstrate value for money, as is the case for the public sector, which must account for expenditure to the government on behalf of the taxpayer.

Variants of the contractor-led approach such as competitive Design and Build, Turn-key or BOT projects can be tendered on a price competitive basis, ideally on a written scope package / brief alone, which precisely defines the requirements to be met by the tendering contractor / provider. The evaluation and selection of proposals is not an easy matter and must be conducted on a clear set of criteria that treats all submitted proposals on an equal basis. It is important to ensure that the process is transparent and can be demonstrated as such to pre-empt any claims of unfair or biased treatment. Competition need not necessarily be on a basis of price level, but can be such that a price is set and the design proposal representing the best value for money or the most interesting solution can be selected.

Should there be a need to select members of the design team individually on a competitive basis, be it the designer, engineer or construction manager, then contractor-led procurement is not a suitable option, although the "develop and construct" form of Design and Build would allow it to some extent. Once more, this type of construct considerably negates the advantages of direct contractor involvement and a substantial amount of time is expended for an individually appointed construction team to familiarise itself and work efficiently. The same has to be said for the individual selection of works contractors on price alone. If a client has a particular need to select some of the works contractors directly, he should be aware that it entails taking on a considerable degree of risk, which cannot easily be transferred. If, after careful analysis, it is in the best interest of the client to select trade / works contractors directly, then a management or perhaps a separate trades approach becomes the preferred option.

In negotiated Design and Build, where there is early and direct contact between the Design and Build contractor and the client, there is some room for both to agree on the selection of some of the important specialist subcontractors in advance and still maintain the overall contractual responsibility of the contractor to the client.

To utilise as much of the early involvement tools as possible in the procurement of construction services, it is necessary to involve the Design and Build contractor early in the project for him to introduce his supply chain from the outset and thus maximise the benefits.

Management

With the choice of contractor-led procurement a client has chosen the simplest of contractual relationships in term of numbers of parties to a project. Even when augmented with the services of a project manager, who should be considered a part of the client organisation, it represents the procurement approach with only a single source of responsibility. If a BOT project is considered, then this represents a single source of responsibility over the entire life-cycle of a building or facility, allowing the client to concentrate all his efforts on his corporate business, of which the building, after all, is often only of secondary importance.

Risk avoidance

This is another aspect that favours the application of a contractor-led route of construction procurement. It is for the client to decide how risks are to be treated, whether to transfer or retain them. As with most other aspects discussed here, when choosing a contractor-led procurement system, it is the early involvement of the entire design and construction team with single responsibility to the client that brings most benefit. It is important for the client to know what he expects from the project in terms of function or operational capability and this needs to be effectively communicated at the outset and laid down in a written scope document or project brief to form the basis of their contractual and working relationship. To ensure that a client's trust in the ability of the contractor to deliver upon its promises is not misplaced, careful selection of a suitable contractor is required. Where a client lacks the expertise a professional project

manager is to be brought in to the process to advise and act on the client's behalf. The
BOT approach offers the greatest degree of risk transfer, including not only design and
construction but also the maintenance and operation of the building, perhaps even the
process itself.

The cost of risk transfer is determined by the market and agreed upon either by
negotiation or a competition of proposals followed by detailed negotiations. However,
in a situation of uncertainty over future developments the transfer of risk to another
party can become prohibitively expensive and certainty of cost and time to completion
difficult to achieve. Such a situation calls for a maximum of flexibility, which favours a
management method, but in its pure form of Construction Management for fee or
"agency" Construction Management, as it is sometimes known. The experienced client
with adequate resources may decide to undertake all activities himself and choose a
separate trades contracting method, which offers the highest degree of flexibility, at the
same the greatest effort and the least degree of risk transfer.

Operation and maintenance

It is a BOT approach to construction procurement, and to a lesser degree Turn-key, that
offer the possibility for maintenance and operation as well as design and construction to
be performed under a single source of responsibility. While it is fairly common to
outsource some facility management functions to a third party, it requires a separate
contract and lacks the synergy of having one organisation design, build, maintain and
operate a building over a long period of time, thus optimising life-cycle costs. While it
is true that this occurs at a price, the enhanced efficiency from utilising a life-cycle
approach should more than compensate the additional cost to a client and he has the
additional benefit of cost and time certainty with payments subject to through-life
performance and risk transfer established.

4.5.4   Some references to German contracting practice

It was already mentioned that some confusion may occur when referring to traditional
contracting in different parts of the world. In the Anglo-American sense, may it be the
United States or the United Kingdom and the Commonwealth (e.g. Australia, Canada,

etc.), it refers to the method of general contracting, where one contractor, based on fully detailed contract documents, is responsible for the carrying out of the construction works. Traditionally, he was performing general building work himself and brought in subcontractors for specialist and fitting-out work. Nowadays, there are very few general contractors having their own directly employed labour force and their main activities are limited to organising the supply chain, co-ordinating construction activities and ensuring that contractual obligations in terms of cost, quality and time are met.

In Germany, however, traditional contracting still refers to what is known in the United Kingdom as separate trades contracting, where the architect is responsible for arranging the supply chain and co-ordinating the work in addition to his duties as designer. The chart below approximately correlates procurement types as described previously in chapter 3.3 with German counterparts[397].

separate trades contracting	=	Einzelunternehmer
traditional, single stage contracting	=	Generalunternehmer - type 1
Develop and Construct D & B (on scheme design basis)	=	Generalunternehmer – type 2
competitive D & B (on a functional or conceptual basis)	=	Generalunternehmer – type 3
competitive or negotiated D & B	=	Totalunternehmer / -übernehmer./ Generalübernehmer
Construction Management for fee	=	Generalmanagement

Table 7: Correlation between procurement types

Generally, it must be said that construction procurement methods are not as clearly defined as they are in the Anglo-American sphere, perhaps, since a single standard contract only exists, which is the "Verdingungsordnung für Bauleistungen" (VOB) in three parts and VOB part B representing the building contract for the majority of construction works in Germany. It attempts in a very concise but general approach to cover all aspects relating to construction works and is considered to be an improvement

---

[397] Gralla, 2001, pp. 81; Kochendörfer and Liebchen, 2001, pp. 58; Sommer, 2000, pp. 9.

on the "Bundesgesetzbuch" (BGB) reference to contract works of any kind. For this reason clients attempt to modify the standard contract, or rather add to it a substantial number of additional clauses in separate volumes known as "Zusätzliche und Besondere Vertragsbedingungen", in order to serve their needs, often in conflict with the law regulating general terms of business transactions called "Allgemeine Geschäftsbedingungen Gesetz" (AGB).

To this day, there exists no other form of standard contract, which could set a precedent and introduce in a standard format different procurement options to both the client and the construction industry, tailored to the needs of the project and providing advantages as described earlier in chapter 3.6.

Finally, it is important to note that German government policy in respect of public sector building works has been one of supporting small and medium sized regionally based construction firms. It is a strict policy to have construction works tendered by separate trades on nearly all public sector projects with all the tender preparation that this approach entails. Only in relatively recent years, for a number of factors including, but not limited to, the re-unification and corresponding lack of in-house capacity, pressures of time to completion and a lack of financial resources, have exceptions to the rule taken place and general contracting and package deals allowed to happen. In respect of BOT style projects, for either infrastructure or public sector buildings, progress has been very slow with only a couple of river crossings under construction in the north of Germany at Lübeck and Rostock and a handful of infrastructure projects being considered. Building projects are at best at a stage of consideration only at the state (Länder) level.

5 The Relationships of a Design and Build Contractor with other Participants

5.1 The Design and Build contractor and clients

5.1.1 General comments concerning the contractor-client relationship

The overriding purpose for a client to undertake a construction project or a number of projects is to improve the effectiveness of his operations and hence service and profits, as was explained previously in section 2.1.5. The objective of both contractor and client can be expressed in terms concerned with the efficiency of the firm such as increasing profitability, improving service, maintaining existing clients and attracting new business. While professional practice / consultants may claim that they are less entrepreneurial than contracting organisations, conflicts, nevertheless, between the needs of individual firms and the needs of the project will arise[398].

Clients generally are not particularly interested in the process to what is being delivered, but prefer to think beyond. Unfortunately, many clients will not fully understand how their facility will perform because the construction industry does not necessarily know itself. Seldom are lessons learnt from one project to the next and it is preferred to perpetuate the notion that buildings are entirely prototypes. It is claimed by clients that advisers must pay more attention to understanding their employers business demands in order to provide the right level of consultancy and so deliver buildings fit for their purpose, since a lack of appreciation of the client's business needs and options leads to design with poor functionality and high maintenance costs[399]. A construction process led by the producer with responsibility not only for design and construction but also for its performance can provide the key to improve effective integration between the client and the supply chain by offering a clearer focus.

There are no hard and fast rules for the level of integration with the client, as much will depend upon the particular views held by the client and his experience of construction projects. However, whatever is devised needs to be clearly communicated and understood by everyone included, particularly by his own organisation. The essence of

---

[398] Walker, 1996, p. 9.
[399] Lamont, 2001 a).

integration is that the decisions made as a result of contact with the client are controlled in terms of the objective of the project. Unilateral decisions made by either the client or the contractor can lead to confusion, which will need considerable unravelling and abortive work, or worse may already be incorporated into the project, with the result that one objective may be satisfied but one or more of the client's other objectives defeated, which in the long run may be more significant to the client's satisfaction with the total project[400].

It is not only imperative from a client's perspective that a firm is able to find suppliers with the highest levels of competence possible, so as to provide any given product or service in the most effective and sufficient way, but also for a contractor to fulfil his objectives to a client. This will be discussed in greater detail in the following chapter. It is in finding the appropriate suppliers which is key to the process that practitioners are able, operationally, to understand the most appropriate type of relationship they require with any suppliers[401]. Should a relationship be arms-length and adversarial, or should it be more collaborative and consensual? How much information ought the practitioner expect from the supplier and how much information is it safe to allow the supplier to have from the buying firm? A competent practitioner will also need to know when it is safe to single source from a supplier, when it is appropriate to undertake joint ventures or when preferred suppliers or marketplace supplier tendering is the most effective way of sourcing a construction project.

Thus, the demands that clients place upon the construction process are frequently complex and uncertain as a reflection of the complexity and uncertainty of the modern world. A concerted solution as the standard answer cannot be expected to solve all problems as complex as these. What is needed is a framework for designing the most appropriate solution under specific circumstances[402] [403].

---

[400] Walker, 1996, p. 108.
[401] Cox and Townsend, 1998, p. 326.
[402] Walker, 1996, p. 9.
[403] See chapter 3 for procurement types and their selection.

At the same time clients themselves are not perfect, which is accepted, if not always admitted, by most clients[404]. This occurs especially when dealing with contractors, as clients are not necessarily able to fully understand and articulate their business needs in terms of construction requirements to their supply chain.

If this is the situation that a contractor is confronted with when attempting to win a contract from a potential client, there is a need to understand how organisations work in order to organise themselves and also how their clients' organisations work, so that they may be in the most advantageous position to interpret and implement their clients' objectives.

Project teams tend to start developing projects assuming that the client has:

- identified the best means of achieving its objective,
- carefully analysed the spatial, technical and performance requirements associated with its objective.

The information provided by the client is therefore frequently accepted without question as the basis for developing the design, however inappropriate it may be. There are many examples of lack of objectivity throughout the world and no matter how effectively resources are applied in devising and executing designs and construction, if they are not achieving realistic objectives, the inevitable result is waste. Having a contractor that has a stake in the performance of a facility obviously would remedy such behaviour or have a loss on his part as a direct consequence of poor performance.

While the lack of an objective evaluation will invariably lead to unrealistic objectives, the internal politics of the client organisation can contribute equally to a lack of objectivity, distortion of objective and potential problems for the Design and Build contractor. It is important that a Design and Build contractor not only finds the client's objective realistic, but also that he has some understanding of the organisational dynamics which brought these forward, since the role of the client in construction cannot be treated as unitary, nor can the events which preceded the decision to build be

---

[404] Lamont, 2001 a).

ignored. The progress of a construction project involves various groups within a client organisation where interests differ and may be in conflict and can only be explained by reference to the past. In order to be reassured in the objective and have knowledge about past processes there needs to be a high level of trust and compatibility between the client and Design and Build contractor. The team leader will need to ask the client many searching questions before the brief is fully developed. It is debatable whether many clients will either be prepared to or will be in a position to satisfactorily answer such questions. As a result many clients' objectives are unsatisfactory and lead to unsatisfactory projects for which the project teams are likely to carry a large part of the blame, if not the responsibility[405].

Propositions about client involvement in the construction process, that ought to be considered by a Design and Build contractor when setting out in a project, include the following[406]:

- Most client systems are very much more complex organisationally then is commonly acknowledged by project teams, in terms of who wants the building who will use it, who approves it, who controls the money, etc.).

- Members of the project team are often impatient of the complexity and insist on dealing with a single client representative, with whom all the internal politics of the client system can be contained.

- Many of the problems causing design changes, delays and difficulties during the construction phase have their origin in the unresolved conflicts within the client organisation and are exacerbated by too early an insistence on an over-simplified client representative function.

- The earliest decisions taken by the client system have more influence over the way the project organisation is formed and its subsequent performance than those taken later.

- These early decisions have their origins in the client's organisational culture, procedures and structures. They are often idiosyncratic, shaped by social and

---

[405] Walker, 1996, pp. 98.
[406] Ibid. p. 100.

political forces as well as by residues of the client's pre-project history on the decision to build.

- The decision to build is a large scale innovation decision with consequences for existing pattern of resource sharing and risk taking in terms of power conflicts and political behaviour within the client organisation.

- These conflicts and behaviour can critically affect the formation, development and subsequent performance of the project organisation (vis-à-vis client / Design and Build contractor integration) which is set up to manage the project and of which the client system is an initiating component.

While in most cases the problems that these propositions describe are unlikely to be resolved, their importance lies in the Design and Build contractor project team knowing of their existence and being prepared to understand and adapt to the dynamics at work to the benefit of the project outcome and its own well being.

If the realisation that establishing an effective relationship with a client by getting to know clients on an individual basis, gaining knowledge of their businesses and future goals and helping a client becoming successful, is to be the ultimate goal of any construction activity, there remains the need for the contractor to convince the client, in order to achieve its own goals. The realisation that a client's priority is achieving success, and not deciding who to award a design or construction contract to, is a key principal in becoming a successful contractor. He is more likely to avoid meeting a potential client, only to demonstrate how well he performs as a construction company, which is for the most part useless information to client. More valuable to the client is a company that provides insight into making its project planning more effective and profitable. A firm should emphasise the value of its services they can provide the client with rather than emphasising pricing issues[407]. The ability to become involved early in the development process of a project and providing input that increases the value of the completed project and the client's profitability is more valuable today then being a low cost contractor only. A step further and offering suggestions to maximise a client's

---

[407] Kubal, Miller and Worth, 2000, p. 23.

investment in physical building requirements can be determining factors that sets a firm apart from its competition.

Knowledge of the client can be used to sell pre-construction services that can provide a competitive advantage for the construction phase and help succeed a contractor by moving the discussion away from commodity pricing issues to one of recognition of professional services. Learning about a client's expectations of a planned project, a firm can ensure that the proposal response clearly differentiates itself from its competitors and improve its chance of success. Such client based knowledge is a reason why many larger construction organisations separate their marketing and acquisition activities by client industry type and have their acquisition teams specialise only in a certain type of industry, such as hospitals, commercial and retail developments, etc, and maintain key account managers to look after major and important repeat clients[408]. Existing clients are the best source of new business and present opportunities beyond work with their own organisation as they are aware of developments in their industry and can help a contractor to learn who might be planning new projects. Additionally, clients can provide information on appropriate contacts within other organisations through the client's business contacts. A referral or personal introduction can immediately open doors and form stronger relationships than might otherwise be possible. At the very least, a satisfied client can be used as a reference to give entry and business from other potential clients[409].

Any construction firm that positions itself as a Design and Build contractor offering pre-construction and possibly operational services must take note of the views of clients and consider those clients that are to be targeted at all times. It is important to remember that trade contractors have other clients and a different philosophy from companies that deliver whole buildings, but it should never be forgotten that they have to be partners striving for a common goal. Too often, the eventual outcome of a project is determined by the worst performing partner and this includes the client[410].

---

[408] Ibid. p. 23; Seddon, 2001.
[409] Kubal, Miller and Worth, 2000, p. 24.
[410] Lampl, 2001.

Obviously, clients vary in many ways, not only in terms of objectives that they seek to satisfy, but also in differences in their experience of the construction process, the importance of the project to their value system and whether they are one-off, casual or repeat clients with a high and regular spend.

Before describing the differences between experienced, repeat clients and one-off, inexperienced clients, and how such a difference tends to affect the relationship with a Design and Build contractor, there is a group of clients that are best avoided. Clients that contractors have to be wary off[411]:

- emphasise price only,
- do not recognise the value of service provided,
- have a litigious history and an overt confrontational attitude,
- create a hostile work environment,
- do not have sufficient financial resources, and
- want to transfer too much risk onto the contractor without appropriate recompense.

It should be mandatory for a firm to investigate and confirm the ability of a client to supply the necessary financial backing to successfully complete the project in mind as early as possible before expending estimating time, pre-construction services and other support requested by the client. Especially clients that emphasise price only will never become repeat clients, since every project they contract is awarded on the basis of low bid and low price. The public sector is a perfect example of a client that firms never invested marketing budgets on because of its dependence on a lowest price selection process, while largely ignoring a contractor's qualification to complete the work. Firms should recognise that potential clients preoccupied with price issues never represent long-term profitable business relationships for a contractor. They do not recognise the value of services provided and treat the project as a mere commodity[412].

The negative impact from just one unsuccessful project can negate the success of several successful projects. The margins in construction are too small to take

---

[411] Kubal, Miller and Worth, 2000, p. 18.
[412] Ibid. pp. 18.

unnecessary risks and successful firms are just as capable of turning away from inappropriate opportunities as they are in closing good ones[413].

5.1.2 Experienced clients and the concept of "partnering"

Clients in the business of construction procurement are usually regarded as experienced if they are in need of a regular requirement for construction work of similar value and content, who can be described as process spenders, in contrast to those clients that are only infrequent purchasers of construction services and therefore known as commodity spenders. Experienced clients are either involved in construction as their primary activity, such as property developers, or are purchasing construction as an important complementary asset to their value system, such as airport authorities, retail outlets or infrastructure operators[414]. It has been claimed that the gap between what are considered to be small occasional clients and those that are generally large regular experienced clients has substantially grown over the years, where small and occasional build clients are not involved in facilitating change within the industry and are not benefiting from initiatives to improve the process to their advantage[415].

Just being a large client with a large regular spend on construction services as one of its important complementary assets does not mean, however, that it is by right an experienced client. Only if it can display a level of procurement competence, that is the ability to know not just one but the full range of relationship management approaches available to buyers and when it is appropriate to use these under specific contingent circumstances, can it rightly be referred to as an experienced client. Whichever approach is chosen, it must be operationalised to achieve more effective leverage of suppliers. In this way, close, collaborative relationships are often used, but only as a means of imposing a more rigorous performance environment on the supplier.

Clients have been able to engineer their procurement improvements by providing an appropriate trade-off that creates a coincidence of interest with their construction suppliers. Large clients with construction spending as a primary or important

---

[413] Ibid. p. 20.
[414] Cox and Townsend, 1998, pp. 337; Tookey, Murray, Hardcastle and Langford, 2001, pp. 21.
[415] Ibid. pp. 21-22.

complementary part of their value system are able to guarantee a regular and high level of demand in return for a willingness by preferred contractors to be prepared to accept a degree of structured control and dominance by the client over their normal way of doing. In such a relationship the contractor / supplier has to be prepared to forgo or reduce the potential for opportunism against the client in return for the buyer's promise of work in the future. With contractors in the climate of the construction industry as it is world-wide, with low margins and few technological reasons to allow suppliers to monopolise the supply market against potential competitors, it is easy to understand that such regular spending clients have been able to attract suppliers in the industry to attend to their needs[416].

If the client organises his negotiation and selection processes professionally to put pressure on the contractor most effectively it can generally exploit its advantageous position. The most effective way for the client to execute pressure on the contractor is for the client to limit the number of contractors who are awarded contracts in such a way that a group of preferred suppliers is created. In such a situation it is the contractor who needs the client more than the client needs any particular contractor[417].

In this kind of relationship it would be surprising if the client needed to rely over much on the enforcement of performance improvements or compliance with performance benchmarks from the contractor, or through the threat of contractual terms and conditions. The power of the relationship is such that the contractor knows that with any disputes it will not be awarded any more work in the future[418].

British Airports Authority (BAA) is a client with an approximate £ 500 million annual construction budget, which now has entered into its second generation framework deals. The process which already has reduced BAA's list of suppliers from12.000 to 1.500 begins with a demanding assessment of its potential suppliers, looking for a leading edge in a commercial sense and in terms of technology with demonstrable continued support. Framework contractors will have to pass an assessment of value and

---

[416] Cox and Townsend, 1998, pp. 337.
[417] Ibid. p. 340.
[418] Ibid. p. 340.

commitment in order to move onto the next year of the ten year agreement of guaranteed work. Four main areas are monitored, including quality of product, management, delivery and price, which involves market testing a random set of components or service levels. If a firm is identified as failing, it is given support with a view of correcting weaknesses for a re-test, but further failure leads to a "managed exit". The threat of extraction means that suppliers have to demonstrate continuous improvement, is to have encouraged dynamism in suppliers and as one framework contractor has stated, the key to working as a team is the framework itself. This compares to a time 14 years ago, when seven publicly owned airports – Heathrow, Gatwick, Stanstead, Edinburgh, Glasgow, Aberdeen and Prestwick – were bundled together and privatised, and all seven airports applied different procurement rules. A key part of BAA's efficiency drive since then has involved standardising procurement including framework contracts for construction suppliers. Being a large, repeat client and continuously developing its procurement competence it can be seen as a prime example of an experienced client, who maintains its own skilled construction staff, primarily for the purpose of project and procurement management. Its preferred route of construction procurement is that of construction management[419].

The process of partnering[420] has been said to be of use even in a single project (project partnering), but real benefits start only to be available when it is based on a long term commitment between contractor and client with a large and regular construction spend (strategic partnering), however, not sufficiently large or frequent, or simply not desired from a strategic point of view, to keep an in-house skilled construction management team over the long term. It should be considered in relation to general procurement and is aimed at integrating the project team including the client. Its focus is behavioural rather than structural as it aims to change the traditionally adversarial relationship between contributors to a construction project, but particularly between client and contractor. The objective is to achieve the project goals by working together constructively rather than by confrontation[421] (10 p.118). Partnering is not unique to the construction industry and it is usually not a legal or contractual obligation, but can have

---

[419] BAA, 2001, pp. 10-11.
[420] see section 2.2.3.
[421] Walker, 1996, p. 118.

an influence on the interpretation of the contract by the courts, should it come to it if the partnering arrangement has turned sour. Partnering agreements do not generally divide responsibility between the partners, since partnering is by its very nature co-operative, but obligations in the contract are essentially divisive, split into those of the contractor and those of the client[422]. The idea is to adopt a process which over time restores trust in business agreements, open once closed areas of communication and allow project team members to once again accept their individual responsibilities with all other team members supporting those responsible. Common to all partnerships projects is the use of a facilitator, who is a trained professional to guide all participants through the process of a structural series of sessions with key participants (client, contractor, principal specialist contractors, users) to develop an atmosphere of working together to achieve their common goals[423].

A partnership agreement[424] to encapsulate the promises just described includes the completion of a series of partnering sessions (adoption of the Value Management Method[425]) that normally include numerous activities[426]:

- individual team members make each other aware of their individual goals and define common objectives,
- to develop a structural programme to determine how to co-operate to reach these goals,
- to establish of a method of accountability, measurement and evaluation to those goals,
- to establish open communication, including complete electronic connectivity, and
- to resolve problems before they become disputes.

Some of the more important aims are to negotiate a reasonable price, ensure greater programme assurance and a smooth running project free of claims[427].

---

[422] Brown, 2001.
[423] Levey, 1999, p. 181.
[424] Standard Form of Contract for Project Partnering by the Association of Consultant Architects, known as PPC 2000, and the New Engineering Contract Partnering Agreement are two such standard forms of partnership agreements.
[425] For more information on Value Management, refer to: Male, et al., 1998 a), b).
[426] Kubal, Miller and Worth, 2000, pp. 12-13.
[427] Seely, 1997, p. 53.

The concept of partnering[428] as a sourcing strategy may be generally applicable to only a small number of large companies with a repeat construction spend. For the remainder, although useful with a minority of strategically important purchases and a very small selection of suppliers, the act of moving the sourcing of a bought-out item from competitive pressure to a single sourced partnership increases both supply risk and profit impact. It is therefore essential that both partners thoroughly understand the implications in terms of costs and benefits and the short and long-term effects of the relationship[429]. For clients, in respect of construction procurement partnerships, it will always be one unusual, specialist sourcing strategy most commonly used by large companies with a substantial and regular investment in construction services[430]. However, it will pose a very interesting question when investigating how important and under which circumstances partnership sourcing is a viable procurement trend for the general contractor's supply management of his suppliers, which will be addressed in the following chapter.

5.1.3   Inexperienced and occasional clients

Those clients which are only infrequent or one-off purchasers of construction services, who are described as naive in respect of knowledge of construction procurement, are not believed to be included in facilitating change within the industry on their own and lack the power or leverage over construction suppliers in order to improve the process to their advantage, are referred to as inexperienced clients. As clients recognise their own relative incompetence and impotence, they have a vested interest in encouraging a highly competitive and fragmented supply base to exist. At least, it is their traditional advisers, principally the architect and project managers as well, who like to take advantage of such a large number of firms to choose from. This is to encourage a situation where there is a multiplicity of apparently interchangeable supply and margins thus keen. Some clients may wish to behave opportunistically in the case of contracts that force the supply chain to take and manage all of the risks inherent in any

---

[428] Also referred to as supplier alliances, partnership sourcing and strategic alliance.
[429] Ramsey, 1996 a).
[430] Ramsey, 1996 b).

construction project[431]. Unfortunately, the results of this type of behaviour all too often ends in frustration and disappoints all involved as described in chapter two.

This is certainly the case for the majority of any particular buyer, but there is a way out of such a situation as just described. Either the buyer relies on a traditional advisor he happens to know or has been recommended to him by some close associate, which, as a rule, is often not a promising way forward, or else, he seeks out a reputable specialist advisor experienced in the field of the proposed project. This frequently is a project manager, who is either an independent specialist or as is often the case nowadays an extension of a multidisciplinary consultant.

Alternatively, of course, if a Design and Build contractor is in the market who has successfully established himself as a reputable and reliable partner in the development of a particular type of project, a client may then feel comfortable enough to approach such a contractor directly and come to an agreement by way of negotiation. Short of such a direct approach he may adopt a competitive selection on the basis of a brief of the anticipated project as described in chapter 4.5. By using the service of a Design and Build contractor directly, a client ties into the supply chain of a frequent and experienced actor in the construction industry, who, if professionally managed, can make best use of his position as a regular employer with a large ongoing spend on a wide range of construction services[432], the benefit of which passing on to the client in some measure.

The marketing activities of a general contractor and especially of a Design and Build contractor must be aimed as high in the client's organisational structure as possible and must be to the decision maker, who has the authority to make selection decisions. However, it is common nowadays to find clients involving consultants in their design and building programmes for a number of reasons and each client structures the use of outside support differently. Clients outsource project management functions because of a lack of internal resources and a limited knowledge of the construction market. It is

---

[431] Cox and Townsend, 1998, pp. 341.
[432] see also chapter 6.2.

important, however, that a Design and Build contractor and his specialist suppliers are involved as early as possible in the project development stage in order to contribute to improvements in the design and construction phases to a higher degree and improve upon its competitive position. A Design and Build contractor has to present its package of services directly to a client and become involved in the early planning stages of a project, rather then merely wait for a consultant to make all the choices. Many clients commission architects because contractors have not informed the client that they can provide early design services, including pre-construction tasks, which eliminate the necessity for a third-party consultant to be commissioned[433].

5.1.4   Public sector clients

Where the public sector client behaves traditionally and maintains large in-house resources to undertake all aspects of project development, design and construction management and in situations where some of these functions have been outsourced among consultants, there simply is no common goal for a relationship to develop between a public sector client and a Design and Build contractor. The analysis has shown, however, that in many countries public sector clients have become aware of alternative forms of procurement and have indeed adopted Design and Build style procurement approaches. The concept of not only contracting out the design and construction phases to a single source but including the finance, maintenance and operation of a facility as described previously is not as widespread, but is well established in a handful of countries, particularly the United Kingdom. A recent NAO publication[434] has confirmed once again that outsourcing the whole aspect of public sector infrastructure provision offers value for money[435] for a number of reasons, not least due to increased efficiencies brought about by single source responsibility and profit motivation to optimise the performance of a facility within the parameters of a concession contract. Such an approach offers incentives for a contractor / provider to seek best possible overall performance from within, as he carries the consequences of poor performance or even failure himself.

---

[433] Kubal, Miller and Worth, 2000, pp. 327.
[434] National Audit Office, 2001.
[435] Value for money gains are defined as improvements in the combination of whole life costs and quality that meet the user's requirements. They will be secured as a result of positive actions by staff involved in commercial transactions. Office of Government Commerce, 2000.

Again, it must be pointed out that BOT is not a procurement method suitable for all of public sector construction and facility management requirements and a public sector taking construction procurement seriously should adopt an approach as described for experienced, repeat clients. This would certainly be possible if the public sector pooled resources and thought about the process holistically, particularly keeping in mind the actual users of a particular building or facility and adopt an approach which is appropriate under the specific circumstances. In order to achieve maximum value for money from taxpayers funds, the public sector should, however, realise its potential as a large and important client and seek to optimise its procurement method, which in the ed will be of benefit to the economy as a whole.

5.2   The Design and Build contractor and consultants

5.2.1   The relationship between contractor, consultants and designers

The traditional separation of powers in a construction project establishes an arena where control of the project is a potential source of conflict. The architect is responsible for design issues but the contractor is largely responsible for all methods and many materials for actual construction. This separation is a factor that denies the intrinsic link between design and construction[436]. Design and construction, however, are extensions of each other, but project participants perceive control of the overall project as being crucial to the achievement of a successful outcome.

The adversarial nature of the relationships between designers and contractors is recognised as one of the most serious problems in the building industry of most countries. It has been suggested[437] that a lack of holistic conceptualisation of the contract and its relation to integrate hampers the ability to achieve improved performance. As referred to earlier, the move towards functional specialisation and professionalism, as the design and construction function separated from each other, has come about in response to the increasing complexity of the construction process. Along with the organisational separation each group of specialisation developed its own unique culture. The design function became a professional occupation (architectural and

---

[436] Puddicombe, 1997, p. 247.
[437] Ibid. p. 245.

engineers) while the actual construction was the province of craftsmen and businessmen[438]. This resulted in an institutionalised, functionally separated, project structure that affects all stages of the design – construction process and is still dominant today. A traditional view, that has been described by one respondent to a survey on this subject, has been that "Architects are the most idealistic and naive, the builders are the most cynical and worldly"[439]. It is a separation that is much stronger than that between functional departments and bringing together these functions with widely divergent cultures is a considerable risk of cultural clash, resulting in a negative impact on the project. The stereotype view of the architect - contractor relationship suggests that designers must act in a manner that protects the client from the contractor, who will operate in a devious and manipulative manner. However, despite these difficulties the increasing complexity and competitive pressures on the industry as described in much detail earlier on indicate that efforts at integration are necessary and will continue, despite evidence that architects and contractors do differ as to the appropriate degree of integration [440]. Architects appear to support a traditional arm's length orientation, while contractors prefer integration, perhaps reflecting a belief on the architects part that they can plan and design for most eventualities.

Thus, the co-ordination of the integration process between design and construction is seen as one of the major areas of difficulties, delays and disputes as to the responsibilities of the parties for resolving design conflicts in the procurement process or for defects after completion. Interface control is a vital aspect of co-ordination. There are four interdependent aspects of co-ordination to be dealt with[441]:

- technical compatibility,
- dimensional integration,
- process planning, and
- flow of information.

---

[438] Probably more pronounced in the Anglo-American sphere of influence.
[439] Ibid.
[440] Ibid.
[441] Hughes, Gray and Murdoch, 1997, pp. 47.

There are few examples of the complete design process being planned and managed as a single integrated process apart from some successful Design and Build projects[442]. The design process is complex and each project generates its own sequences and priorities. Effective design management requires flexibility and an understanding of both the process and the contributions required from the people involved.

An integrated approach to design and a comprehensive management approach to that design and the elements of co-ordination is clearly a pre-requisite not only to the effective engagement of specialist contractors on any project, but also for the successful completion of any project to the satisfaction of the client and of advantage to the contractor[443].

A Design and Build contractor is ideally placed to ensure that a comprehensive management framework is established at the outset to facilitate the proper integration of inputs from all the specialist contractors into the design process. Thus a complete design is generated for the satisfaction of needs of ever increasing complexity and component based construction.

5.2.2   Good practices for the relationship between contractor and consultants

Good practice for a Design and Build contractor requires that the design manager has the authority to make decisions about the contributors to the design and that the design team have clearly carried their own design forward to a point where each specialist contractor can be effectively brought into the process. Rather than the sequence of construction it is the sequence of design which governs the decisions for timing the appointment of contributors and specialist contractors, thus reversing the traditional priorities[444] [445].

Design and Build is the ideal contracting format for the participation of specialist contractors early in the design process. Subcontractors, who are experts in their

---

[442] See examples in section 4.5.2.
[443] Ibid. p. 80.
[444] Ibid. p. 46.
[445] see also 6.3.5.

particular construction specialism, can provide valuable insight into constructability issues for the overall advantage of a project proposal and work alongside with the designers as part of the project team[446]. The principal objectives of early involvement are effective cost management and improved functionality leading to a better value construction project. The greatest benefit from early involvement is obtained before construction starts on site. Such early involvement stops thinking of the construction process as a series of sequential stages and encourages the adoption of a concurrent, holistic approach to briefing, design and construction[447]. For the design team it will mean:

- working more closely with contractors and specialist suppliers than they have been used to in the past,
- overcoming attitudes relating to designers' pre-eminence in the supply chain, and
- devolving responsibility for detailed design to those most able to provide competent solutions.

A design manger's or Design and Build co-ordinator's key duty is to liase between the bid team and designers to ensure all the current design and documentation for a project are made available, liase between specialist contractors and the design team. Dependent on the nature of the project, many of the specialist design duties are undertaken by specialist contractors, which still need co-ordination from a specialist engineer / designer and the duties undertaken will be on instruction from the Design and Build co-ordinator, who will police the tender submission documentation to the client[448].

In a Design and Build contracting format, the quantity surveyor's (QS) duties are fundamentally taken on board by the Design and Build contractor, where a bill of quantities to a relevant method of measurement needs to be prepared to enable estimates and any subcontractor or supplier to accurately price the proposed project. Subcontract packages are identified, procurement routes selected and the issue of enquiry documents and analysis of returned quotes undertaken. When the design is completed to outline / scheme detail during the bid stage, the QS has to make assumptions to complete the bill

---

[446] Kubal, Miller and Worth, 2000, p. 349.
[447] Hill, 2000, p. 4.
[448] Simm, 2000.

of quantities in the short tender period given by some clients. Advice will be given on the economies of the design during completion and alternatives will be proposed for consideration by the bid team. Other duties are the preparation of stage payment charts, cash flow forecasts, principal quantities schedules for tender submission, the preparation of activity schedules and cost component schedules. In return the requirements of the on-site QS are reduced as there are no re-measurements and additional payment / claims duties, unless variations are introduced by the client. The assumption of an internal financial role becomes his core function[449].

A Design and Build contractor will need highly developed project management skills, as there may be a reluctance on the part of members of professional practices to be managed by construction companies, unless the practices are carefully selected. The client may also retain a project manager and other professional advisors to oversee the Design and Build contractor. If professional skills are all in-house to the Design and Build contractor the relationships with the client and client's advisors should not be too difficult to manage[450]. A Design and Build organisation structure should reduce differentiation and provide a sound platform for effective integration resulting in a proficient management structure. However, in practice it is not very often that all the project skills are in-house to the Design and Build contractor. Frequently Design and Build companies do not have all professional skills in-house for a number of good reasons, not least because of the decision to limit investment and the problem of retaining the scope of skills which may be required. As a result professional skills are hired-in from individual professional practices in a number of ways[451].

5.2.3   Alternative approaches for design completion

Such arrangements create principal challenges to the designer-contractor interface, including:

- how to create an incentive for the designer to generate a good value for money solution not only for initial but for through life costs as well,
- how to obtain alignment of the goals of the designer and the contractor, and

---

[449] Ibid.
[450] Walker, 1996, p. 116.
[451] Ibid. p. 116.

- how to generate competition between design teams without incurring prohibitively high costs.

The Design and Build contractor is generally lead by the contractor, yet the group that has the greatest influence on costs are the designers. The designers normally shoulder little risks. Their contractual relationship with the contractor can be no different from the one he enjoys with the client. Worse, if fierce competition has invaded the designer-contractor relationship too, the designer has every incentive to keep his own costs to a minimum but, once the contract is awarded, none whatsoever to help the contractor to reduce his costs[452], if the designer is not interested in an ongoing relationship and does not want repeat business.

One solution to the problem is for the Design and Build contractor to employ his own designers directly. A second is to develop partnering or alliancing arrangements and a third is to develop risk and reward sharing schemes. A further possible method involves the designers to lead the Design and Build consortium.

Most general contractors do not have in-house architectural and engineering capabilities worth mentioning. A firm has to either purchase a firm with these capabilities, hire a complete staff of architects and engineers or form one or more alliances with architectural and engineering firms to jointly offer Design and Build services. The latter is typically the most common and realistic option to provide Design and Build services for most general construction[453].

Problems in retaining in-house staff include:
- not being able to design a wide range of project types,
- not having sufficient knowledge of all possible geographical locations,
- not being in a position to provide a continuous and uniform rate of working, and
- the inability to provide excellence or a certain reputation in all sectors of work.

---

[452] Horner, 1999.
[453] Kubal, Miller and Worth, 2000, pp. 339.

For most Design and Build contractors outsourcing architectural and engineering work is the only option available to complete Design and Build work. They can execute the design process either through an alliance or by simply subcontracting the design to a suitable practice.

An alliance with selected architectural and engineering firms is the most effective way to maintain Design and Build capabilities and to market Design and Build projects. They are normally structured on an exchange basis for a particular project sector, geographical area or a combination of both. Often a Design and Build contractor will team with several design and engineering firms based on expertise and geographical location, ensuring that none of the relationships presents a conflict of interest with other teams. In the same manner architects and engineers are expected to maintain connections with multiple contractors based on similar criteria[454].

A Design and Build contractor can subcontract the design package to an architectural or engineering firm in much the same manner as the actual site work is subcontracted, bearing in mind all the advantages and disadvantages of a range of subcontract approaches to be described in chapter six. It allows the Design and Build contractor the greatest choice of design and engineering firms and theoretically it can select what it feels to be the most appropriate design firm for a particular project without having to form an alliance and having to pre-select and thereby limit itself. This is often the choice of large contractors, who can bring sufficient leverage to bear on the market and are attractive for their ability to offer repeat business, and market a wide variety of project types. In the United States it is the subcontracting of design which is probably the method frequently used by most Design and Build firms[455]. Subcontracting the design allows a contractor using its preferred method of working without commitment to a particular architect and permits him to identify and choose the most appropriate design firm for a specific project, thus offers the greatest degree of flexibility.

---

[454] Ibid. pp. 339.
[455] Ibid. p. 340.

However, the most realistic practice is to form a teaming relationship or project partnership based on intelligence gathered before a proposal is released and entering an agreement with a design team before the proposal / development process begins. Thereafter, the contractor may find that the most experienced and suitable design firms for the required work have already been committed to other contractors[456].

A contracting firm with internal design and construction capabilities is suited for work in specific niche markets such as housing, pre-fabricated building systems or specialist process facilities rather than an attempt to maintain resources necessary for a wide range of design and build capabilities. Even firms that maintain internal design resources will typically not maintain a full complement of skills that would have to include architects, structural engineers, mechanical and electrical engineers, landscaping and interior designers and so forth. Contractors that claim to be true Design and Build firms often have to outsource major portions of a project's design to other design firms or teams as would any other contractor in order to provide the design expertise required[457].

Each type of design completion method, whether they are subcontracting, alliancing or in-house design resources, has its advantages and disadvantages that a Design and Build contractor needs to be aware of and are summarised in the table over the page.

---

[456] Ibid. p. 340.
[457] Ibid. p. 341.

type of design completion method	advantages	disadvantages
subcontracting	-permits the best design team to be chosen for the specific project without having to depend on pre-selected or in-house capabilities. -the contractor can promote Design and Build without regard to any pre-established alliances or internal capabilities.	-clients may perceive that the contractor is not a true Design and Build contractor if it does not have design capabilities. -the most appropriate design team may have already committed itself to another contractor or alliance. -full integration of the designer into the project team may be difficult to achieve.
alliance, strategic partnering	-the contractor can emphasise the ability to bring in just-in-time talent for the design and construction phases to increase cost effectiveness for the client. -marketing presence is increased for all alliance members by the promotion of the alliance Design and Build capabilities. -the contractor potentially creates the most beneficial combination of flexibility and integration of team members.	-the client may perceive that the alliance is limited to a particular niche targeted by the alliance marketing and reference projects. -formal alliances /strategic partnering agreements can create conflicts of interest when team members attempt to form other alliances or want to work individually if a good opportunity arises.
internal resources	-clients are able to truly deal with a firm for design and construction. -the contractor's capability and track record is readily apparent. -integration of all contributors to the project team should be easiest.	-internal resources are limited in work type and skill and need to be supported by external resources, creating a conflict of interest within the firm. -clients may perceive that the contractor's design team can be pressured by the construction team to reduce costs by lowering the quality of design and that the project team is biased towards construction.

Table 8: Advantages and disadvantages of design completion methods[458]

A proven capability to complete Design and Build projects is more important to a client than having the internal resources or an alliance structure necessary to complete the design work. A Design and Build contractor with a proven track record of co-operating successfully with a variety of architectural and engineering firms to complete a wide range of building projects has the advantage of appealing to a variety of clients. As circumstances of every construction project tend to be unique to some degree, a contractor able to co-operate successfully with numerous design firms may actually be

---

[458] also: Kubal, Miller and Worth, 2000, pp. 343.

in a more advantageous situation than a company with internal resources that operates with a limited number of niche capabilities.

Whichever way a Design and Build contractor chooses to complete the design required, the selection of external design resources should additionally consider the following factors[459]:

- task performance[460],
- contextual performance[461],
- fees,
- and relationship factors.

They combine in such a way that they influence either consciously or intuitively a Design and Build contractor's selection decision as illustrated below.

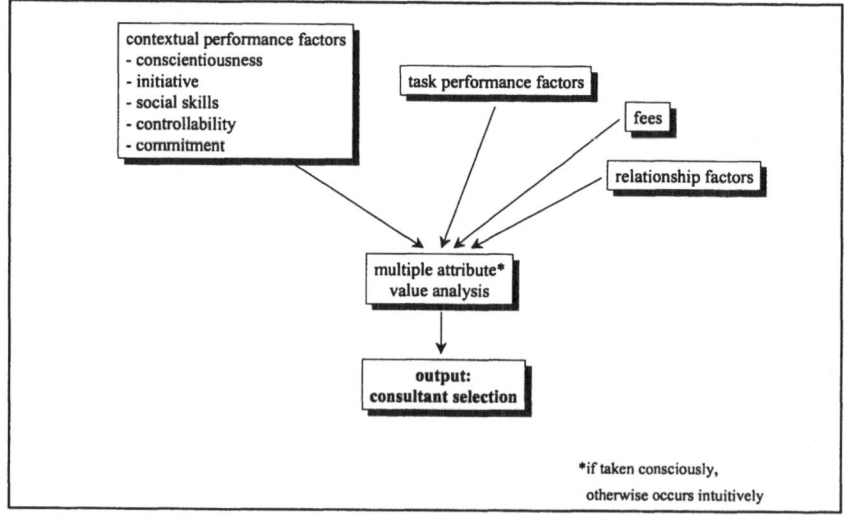

Figure 15: Framework for the selection of consultants by a Design and Build contractor[462]

---

[459] Ling, Ofuri and Lam, 2000.

[460] Task performance is the proficiency and skill in job-specific tasks.

[461] Contextual activities arise because the consultants interact in an organisational setting instead of working by themselves, and, therefore, need to communicate with one another, co-ordinate, follow instructions and occasionally go beyond their job descriptions. Excellent contextual performance occurs when consultants have the appropriate soft skills.

[462] Ibid.

5.3   The nature of contractor to contractor relationships

5.3.1   Types of contractor relationships

From a Design and Build contractor's perspective, there are essentially three different approaches to assembling the required resources, skills and capabilities to undertake all construction activities necessary in completing a project. Much the same as completing the design a Design and Build contractor has the option to perform all works in-house, which is unrealistic for reasons already explained and not a competitive or even sustainable method of working in today's economic climate. Depending on the strategy of the firm, two other approaches are adopted, including a coalition of two or more firms, also referred to as partnering, alliancing or joint-venturing and alternatively subcontracting, where an individual or organisation is employed by a contractor to construct part of a project. Very often, a combination of both is found in practice and a joint venture between two or more firms and subcontracting coexist side by side.

Joint venture contracts are formed for the reasons of limitation of risks, pooling of risks, exploiting opportunities and harmonisation of the whole operation[463]. It is a type of coalition, alliance or partnering and is used for a firm to pursue the benefits of a broader scope with independent firms. Coalitions are longer term agreements among firms that go beyond normal market transactions but fall short of outright mergers. Other forms of coalition, besides joint ventures, are technology licenses, supply agreements and marketing agreements[464]. Coalitions can allow sharing of activities without the need to enter new industry segments, geographic areas or related industries. They are also a means of gaining the cost or differentiation advantages of vertical linkages without actual integration, but overcome the difficulties of co-ordination among purely independent firms.

A coalition in the form of partnering is one of the methods advocated to generally improve the performance of the construction industry. However, it is large construction companies which are expected to enter into such arrangements with clients and be at the

---

[463] Seely, 1997, pp. 89-90.
[464] Porter, 1985, pp. 57.

forefront of changes to improve productivity. Construction SME's[465] on the other hand are expected to join partnering relationships instigated by large main contractors[466]. This topic will be investigated in closer detail in the subsequent chapter. Meanwhile, it is SME's who comprise the bulk of the construction industry and are well positioned to take advantage of new market opportunities arising from collaborative building programmes, but are not without many difficulties.

Difficulties in reaching coalition agreements and in ongoing co-ordination among partners may break a partnership or nullify the benefit. Partnering firms remain independent firms and there is the question of how the benefits of a coalition are to be divided. The relative bargaining power of each coalition partner is thus central to how gains are shared and determine impact of a coalition on a firm's competitive advantage[467].

Reasons currently named for undertaking a formalised joint venture (ARGE) under German law, are[468]:

- Smoothing of the work load, where frequent co-operation in joint ventures allows a more even and continuous use of resources, which is dependent on progress and subject to unavoidable changes during the course of a construction project, otherwise not possible as a stand-alone company.

- Spreading of risk, where a firm no longer bears the full risk which is limited to the share of the individual company in the joint venture. There are, however, different ways of allocating risk within an ARGE joint venture and the joint venture will always be jointly liable in respect of the client.

---

[465] A Small Medium Enterprise (SME) is an enterprise which has fewer than 250 employees, has either an annual turnover not exceeding 40 million Euro or an annual balance sheet total not exceeding 27 million Euro and conforms to the criteria of independence. These contain further stipulates that not more than 25 % of a SME may be owned, either singly or jointly, by a large company. A small company must have less than 50 employees, an annual turnover not exceeding 7 million Euro or an annual balance sheet total not exceeding 5 million Euro. Companies with less than 10 employees will be considered "very small". Cordis RTD, 2000.

[466] Davey, Lowe and Duff, 2001, p. 42.

[467] Ernzen and Schnexnayder, 2000, p. 57.

[468] Burchardt, 2001, pp. 858.

- Improved access to large and prestigious projects, which allows the individual company to partake in carrying out projects that are large, complex and may be prestigious.

- Transfer of expertise and marketing of specialist skills, where opportunities for a specialist company are widened and has a chance to become involved in a greater number of projects, or a company gains expertise improving its core capabilities.

- Avoidance of subcontracting, where the joint venture helps to reduce the use of "unreliable" subcontractors, or from a SME's perspective a subcontracting role can be avoided to contract directly with a client as part of a joint venture. It is unlikely, however, that subcontracting will be avoided altogether, nor is it desirable to do so.

- Co-operation versus competition, where firms are free to form a joint venture, without compromising competition law, if they are jointly of the opinion that it is in the best interest of all parties concerned and will be advantageous in commercial terms. It provides the opportunity to pool resources and responsibilities if independent firms form a joint venture in good time to bid for and later carry out the works.

- Increased flexibility in reporting requirements, where the balance of a joint venture project is included under current assets on reporting day, while individual projects have to be reported in the profit and loss account.

There are generally three types of construction joint ventures defined within German law[469]:

(1) Bietergemeinschaften (bidding joint venture / pre-contractual co-operation), where the coming together of two or more firms to jointly bid for a project is contractually defined. Nonetheless, the bidding joint venture will stop to operate if either the bid was unsuccessful or successful. If successful, members will undertake the project based on the conditions already stated in the terms and conditions of the bidding joint venture contract.

(2) Dach- / Los-ARGE (umbrella or package joint venture), where the work under the contract is separated into trade or elemental packages, which are allocated among

---

[469] Ibid. pp. 859.

members to the joint venture in the form of a subcontract. All members are jointly liable in respect of the client, however, individually responsible for their package only in respect of the Dach-ARGE company (joint-several liability).

(3) Arbeitsgemeinschaft (ARGE / joint venture), which is a coming together of two or more independent companies to form a joint venture company with the aim to fulfil the obligations under the contract with the client. They are jointly liable for the outcome of the joint venture enterprise.

All joint ventures of the type described above are usually entered into on the basis of standard contracts issued by the Hauptverband der Deutschen Bauindustrie (HDB)[470], the Central German Contractor's Association.

German construction companies, including the largest firms, often formed such joint ventures companies, which were usually made up of firms of similar size and structure and served to limit competition and pool resources for the completion of a particular project. All losses or profits were shared on an equal basis and the client benefited from joint liability of the joint venture members. The subcontract content was limited to less than 50 % of all activities and restricted to specialist work. More progressive firms are now prepared to do the work independently and the subcontract content has grown to between 70 % to 80 % of all activities. Only with large projects are joint venture solutions sought, however frequently only with clearly defined liability and responsibility over specific work packages only (Los-ARGE) and not jointly for all activities[471].

While it is a valid argument for SME's to enter into a coalition of one type or another to improve upon its competitive and bargaining position, including benefits of economies of scale, learning, access to new markets, needed technologies or to meet client and / or government requirements and to spread risks, it is not such a valid argument any longer when considering large contracting organisations[472].

---

[470] www.bauindustrie.de.
[471] Mehrtens, 1996.
[472] refer also to: Porter, 1990, pp. 66.

Coalitions carry substantial costs in strategic and organisational terms. The very real problems of co-ordinating with independent partners, who often have different and conflicting objectives are just a start. Co-ordinating difficulties impede the ability to gain the benefits of a wider strategy. Today's partners often become tomorrow's competitors. Coalitions or alliances are unstable and are frequently transitional devices. They proliferate in industries undergoing structural change or escalating competition, where managers fear that they cannot cope. They are a response to uncertainty, and provide comfort that the firm is taking action. In the long term, global leaders, if ever, rely on a partner for assets and skills essential to competitive advantage in their industry. The most successful coalitions are highly specific in character, which are narrow in focus and orientated towards access to a particular market or technology[473]. Coalitions are a tool for extending or reinforcing competitive advantage, but rarely a sustainable means for creating it[474].

5.3.2 Specialist contractors and subcontractors

The reasons for subcontracting not only to exist but also to continue spreading, were previously discussed in detail. It is sufficient to say that main contractors will require subcontractors of high calibre and with appropriate resources to execute the necessary works at a price and quality that will enable main contractors to be competitive in their overall tender to the client. Any selection and tender process requires fair dealings between partners as a basis for successful teamwork and the avoidance of disputes[475].

A recent empirical survey in the United States[476] revealed the following reasons for subcontracting to take place, which are ranked in order of importance as follows:

(1) need for reducing liability exposure

(2) reduce overhead costs

(3) reduce overall construction costs

(4) market volatility

(5) faster construction time

---

[473] Very good examples for this type of coalition are joint ventures between contractors and operators forming a Special Purpose Vehicle (SPV) for operating BOT infrastructure facilities.
[474] Ibid. pp. 66.
[475] Construction Industry Board, 1997, p. 7.
[476] Costantino, Pietroforte and Hamill, 2001.

(6) reduce equipment / maintenance costs

(7) better value to the client

(8) better workmanship

This ranking cannot represent every construction market everywhere, but it offers a range of reasons which taken as a whole explain the existence of subcontracting. Especially the argument of unstable market conditions is commonly put forward as the overriding reason for general contractors to transact with subcontractors, since it enables them to be flexible in responding to potential market up and downs[477].

It is necessary at this point to clearly define what is actually meant when referring to subcontractors. "Subcontractor" refers to those firms with responsibility for some part of the construction work (whether with or without design service) under the employ of a main contractor. Often the term is also used to cover those firms with a subsidiary relationship e.g. works contractors under the employ of a management contractor. Such a relationship is not as a matter of course an indication of respective sizes or bargaining strength of either contractor or subcontractor, however, it is frequently the case that a contractor is both larger and in a better bargaining position than a subcontractor. A "specialist trade contractor" is a generic term for firms who offer and execute a specialism in any or all design, manufacture, production, assembly, installation, testing and commissioning of items that go into the construction of a building. Specialist trade contractors have three different origins[478]:

- the practice by main contractors of subcontracting the labour content of the work[479],

- the emergence of trade contractors who have replaced the main contractors' directly employed craft operatives, and

- the proliferation of technologically advanced firms.

---

[477] Kale and Arditi, 2001.

[478] Hughes, Gray and Murdoch, 1997, pp. 10.

[479] However, even in the United States little, if any, labour only subcontracting exists nowadays; Costantino, Pietroforte and Hamill, 2001. This incomplete form of contracting, although permitted in many countries but not generally in Germany, does not allow a clear cut transfer of responsibilities. Quality problems and claims occur at the interface between the supply and installation of components and materials. This possibility is avoided by main contractors in allowing full subcontracting to transfer risk and liability and also to manage the complexity of construction technology.

Of course, the sub-letting does not necessarily stop after a main contractor has sub-let a work package to a specialist trade contractor. There are also "sub-subcontractors", an individual or organisation employed by a subcontractor to construct (and sometimes design) part of a project[480].

Particularly among specialist contractors, who provide a full service from design to installation on site utilising specialist skills and equipment, can some be of considerable size operating on a global scale within their niche. Typical sectors include ground works engineering, the mechanical and electrical and building facade sectors.

It is recognised that for the construction industry to improve its productivity, capability and cost effectiveness, the competence of subcontractors must be enhanced[481] on the one hand, and on the other hand main contractors have a responsibility to align the resources of their subcontractors and suppliers to meet the needs of the client[482].

This is not surprising, since a typical main contractor's overhead or purchasing item of a project take up about 20 % of the total construction costs, while 80 % is represented by materials and services costs contributed by sub contractor and suppliers[483]. Hence a company depends very much on the co-operation of the subcontractors and suppliers in controlling or cutting their costs. Moreover, since most of the work of a project is done or provided by subcontractors and suppliers the nature of co-operation significantly affects the progress and quality of the project they handle[484].

The characteristics of main contractor – subcontractor transactions of high asset specifity and uncertainty[485] coupled with specific quality objectives, budget restrictions and time constraints present numerous challenges to the parties involved in construction. A main contractor can partly address those challenges by establishing and

---

[480] Construction Industry Board, 1997, p. 4.
[481] Loh and Ofori, 2000.
[482] Wong and Fung, 1999.
[483] Ibid.
[484] A hands-on account of specialist trade contractors in their role of subcontractor highlighting the need of positive co-operation, their significant input into a project and many of the difficulties experienced in their dealings with main contractors is given in: Building 26/10/2001.
[485] Kale and Arditi, 2001.

maintaining good relationships with subcontractors, since relationships of high quality facilitate the function of subsequent transactions. In these subsequent transactions parties can rely on the experience acquired in previous transactions to overcome problems of commonalties, communication and integration. Any transaction in which the performance of the two parties is separated by time involves an element of trust. The stability of the relationship is associated with the investment in trust by the parties concerned. Strong ties, indicated by the length of the relationship between entities in project based industries, have three basic characteristics[486]:

- frequent interaction
- an established history
- mutual intimacy or mutual confiding

They are likely to promote long term connections and facilitate information exchanges. On the other hand, firms seek competitive as well as co-operative advantages. Firms exhibit rivalrous behaviour, erecting barriers and thus distinctive areas of competence.

5.3.3   Current nature of main contractor-subcontractor relationships

While the use of subcontracting is widespread and continuous to spread, which supports the view that related transaction costs are lower than in-house resources, even if such lower transaction costs are expected and not measured in practice, and justifies subcontracting[487], it is reported that an "adversarial culture" sometimes exists between main contractors and subcontractors and that this can lead to a poorer project performance[488]. Projects, where the subcontractors' impact has been badly managed, can generate antagonism between the parties and cause serious contractual disputes. Antagonism may be the result of one party not performing properly, usually through failing to understand or acknowledge the needs and objectives of others in the project. In many cases main contractors invite tenders from subcontractors at a stage when they themselves are not yet and may never be appointed to undertake the work. In these circumstances the time available for tendering and the information that can be provided to tenderers is often not in the direct control of the contractor. If a contractor is given

---

[486] Sözen and Kayahan, 2001.
[487] Costantino, Pietroforte and Hamill, 2001.
[488] Hughes, Gray and Murdoch, 1997, p. 17.

insufficient time or information by a client for the preparation of tenders, the effectiveness of the subcontractor selection process will suffer[489].

Subcontractors are known to consistently sign contracts anticipating that none of its highly restrictive clauses will ever be invoked. In fact, if they read the subcontract agreement "word-by-word" in the presence of their lawyer, it is doubtful that many such contracts would ever be signed. A look at some of the more onerous provisions in the standard main contractor's subcontract agreement will prove the point[490] and should impress upon the subcontractor the need to read and understand the provisions of their subcontract document[491], regardless of the varied forms of protection offered by individual national laws in respect of lawful or unlawful terms and conditions of contract. Such terms and conditions of contract often include for example: pay when paid clauses, binding subcontracts to all contract documents, agreement of receipt of complete drawings, specifications, addendum's, etc, articles dealing with "intent" of the contract documentation, directive to work clauses, "perform or else", onerous termination and compensation clauses and restrictive arbitration / adjudication clauses.

The increase in complexity, the oversupply of specialist firms and the declining construction output in many markets has cultivated an adversarial atmosphere, which has a negative effect on the main contractor – subcontractor relationship. As main contractors have realised that the greatest potential for cost savings lies with subcontractors (80 % of total project costs), the extent of unfair contract conditions, bid shopping and other onerous practices has increased. Subcontractors have also caused problems. With easy entry into the construction market place, subcontracting organisations have been established with very little capital investment. Often, subcontractors do not have the necessary expertise or resources to undertake the work satisfactorily and as a consequence are unable to give their employers the service they require. Many of the bad traits common to main contractor – subcontractor relationships are also common to subcontractor – sub-subcontractor relationships[492].

---

[489] Construction Industry Board, 1997, p. 8.
[490] e.g. Franks, 1997; 1998; Kniffka, pp. 46-65, Hofmann, pp. 66-75, Medicus, pp. 76-85, 1992; Passarge and Warner, 2001; Ruckteschler, 1988; Schwarz, 1996.
[491] Levey, 1999, p. 39.
[492] Kumaraswamy and Matthews, 2000.

Although recent publicity in some markets, particularly the United Kingdom, show a shift in the attitude of main contractors to subcontract procurement, a survey of the specialist contractors' sector showed that this impression should be approached with caution: the typical contractor – subcontractor relationship is still traditional, cost-driven and potentially adversarial. Nevertheless, the two approaches are co-existing, which is consistent with the institutional theory of organisational strategy[493].

Our own survey of relationships between main contractors and subcontractors both in Germany and England[494] has shown that main contractors are "keen" to improve their working practices with subcontractors and attempt to address some of the common problems usually encountered by introducing, for example, formalised start-up meetings and keeping organised and systematic records on subcontractors' performance. However, problems of poor communication, lack of information on site, inadequate supervision, failure to complete on time, incorrect pricing, insufficient quality or wrong products were experienced by all companies included in the survey.

Differences between Germany and the United Kingdom were revealed, in that German specialist trade contractors reported to be in the least preferred role of subcontractor in only 10 % to 50 % of projects compared to 70 % to 90 % in the case of UK specialist trade contractors. In turn, German specialist trade contractors were successful in securing a larger share of direct contracts with clients, in the order of 50 % to 70 % of all contracts and have themselves subcontract 10 % to 30 % of their workload, whereas UK specialist trade contractors only achieved a rate of 10 % to 30 % of direct contracts as a proportion of all contracts. One explanation is the predominance of single-trade letting of the German public sector, which is generally mandatory and thus focuses on smaller, regionally based specialist trade contractors[495].

Nevertheless, a unanimous response was received from all specialist trade contractors to the effect that getting paid still presented the most serious of problems in building up a better relationship over the long term with a main contractor. Main contractors were

---

[493] Greenwood, 2001.
[494] Winter and Preece, 2000.
[495] see also section 4.5.4.

lacking in trust and were overtly suspicious in all their business transactions with subcontractors. Pressure was repeatedly applied to reduce prices and at the same time critical information was held back, making it almost impossible to allow for proper pricing and working. Late orders and not allowing for sufficient time in both the preparation and execution of a work package often created problems in the provision of products and services.

It is not surprising that such problems were all too frequently reported, since the approach adopted in selecting subcontractors is largely based on price alone, on average six quotations were required, and the responsibility to maintain a constructive working relationship is all too often left to the subcontractor only.

It has been reported that in the United States main contractors and subcontractors jointly restrict access to their transactions, whereby main contractors tend to rely on a few subcontractors in each trade to establish long term relationships with these, and similarly subcontractors mostly prefer to work with a rather smaller set of main contractors with whom they establish long term flexible relationships. Problems of newness are overcome by learning from one another, which renders transactions between them highly asset specific. Subcontractors are reported not to submit quotations to those main contractors that have a reputation of bid shopping and are selective in getting involved with main contractors. Similar quotations are submitted to main contractors with whom they have satisfactorily done work in the past and they increase the price submitted to main contractors with whom they have limited experience by 5 % to 10 %[496].

A survey of main contractors in the commercial building construction market in the United States revealed that this type of construction is characterised by considerable contracting out at a rate of approximately 76 % in 1997, with an average number of 10.2 subcontractors engaged for a trade by a main contractor, with a minimum number of 7.2 for vertical transportation and a maximum number of 15.1 for interior finishes and

---

[496] Kale and Arditi, 2001.

partitions[497]. This demonstrates how commercial construction is governed by strong market subcontract conditions and reflects upon the purchasing strategy of main contractors who, depending on their situation, develop more than one package for the same category of work by balancing the need for increased competition with, at the same time, the need for decreased responsibility and supervisory effort. According to the survey the average length of long relationships was 21 years, while the length of a typical business relationship averages 9.6 years. These levels of "fidelity" index show that main contractors maintain business relationships with a select group of subcontractors for a long period of time and proves, at least on average, that it must be beneficial for both parties to have continued for so many years.

The survey additionally showed that the contractual relationships between main contractor and subcontractors is strongly reliant on the type of relationship between client and main contractor in a given project. Negotiated contracts with the client favour a type of business relationship of a main contractor with a subcontractor that is closer to that of a quasi firm, reflecting a thrust towards a semi-integrated form of organisation. Competitive contracts, however, favour more market driven types of relationships between main contractor and subcontractor as demonstrated by the table below[498].

main contractor's procedure for selecting subcontractors	main contractor has a negotiated contract with the client	main contractor has a competitive contract with the client
lowest bidder	4.5 %	9.5 %
lowest negotiated price	24.8 %	52 %
best price from a proven subcontractor	60.2 %	29.5 %
sharing work to maintain business relationship with subcontractor	10.0 %	8.5 %
other	0.5 %	0.5 %

Table 9: Correlation between subcontractor selection procedures and type of main contract

The organisational choice of production appears to follow a thrust towards recurrent co-operation with a limited number of subcontractors that offer competitive prices. The choice, however, is strongly influenced by the extent of market competition that is

---

[497] Costantino, Pietroforte and Hamill, 2001.
[498] Ibid.

experienced by contractors and the type of relationship between client and main contractor[499].

Taking a closer look at the nature of the main contractor – subcontractor relationship from a subcontractor's perspective in the United States, reveals how the majority of subcontractors are of the opinion that bid invitations which they receive from main contractors are generally poor and contain misleading and insufficient information. The following long list of statements expressed by subcontractors illustrates this[500]:

- Invitations provide little information about jobs to be subcontracted.
- Invitations vary greatly from one contractor to another (for the same project).
- Invitations are rated fair to poor.
- Invitations sometimes are misleading and it is hard to locate certain items because of the way that they are arranged.
- The job schedule (programme) is rarely included in the instructions. This can be crucial when deciding on what to bid. Invitations should always include the name, location of bid, square footage and the bid date.
- Invitations to bid do not provide enough information on the size of subcontract work.
- Invitations provide no information regarding "pre-bid conferences" and "walk-throughs".
- Invitations usually need clarification.
- The contracts are adequate for the purpose.
- The invitations are usually complete, basically acceptable, sufficient and adequate.

Following submission of a subcontract bid, subcontractors are notified in the majority of cases only after the award of the project contract that their bid was used and marks the point at which negotiations usually start. The main contractor discusses the subcontractor's experience, current workload, financial capacity and other factors including variations in the work package. The majority of subcontractors negotiate prices only after the project contract was awarded to a main contractor and some abuse

---

[499] Ibid.
[500] Sash, 1998.

the situation for squeezing the subcontractor to reduce the submitted prices in an effort to increase margins[501].

Another survey, undertaken within the Constructors' Liaison Group in the United Kingdom, which represents most of the industry's population of specialist trade contractors and included approximately 700 firms in the years 1999 and 2000, compared the results with the recommendations of the "Code of Practice for the Selection of Subcontractors"[502].

issue	code principles	survey findings
preliminary enquiry	recommended	13 % of tenders
number of competitors	max. 6 contractors	5 competitors
tender prices	min. 6 weeks	usually (91 %) < 3 weeks
selection other than price	recommended	7 % of tenders
selection criteria indicated	recommended	11 % of tenders
conditions of contract	should be indicated	86 % of tenders
start and finish dates	should be indicated	42 % of tenders
payment terms	should be indicated	81 % of tenders
industry standard form	preferred	49 % of tenders

Table10 : Comparison between results and the recommendations of the code

Non compliance with the code was definitely identified in six out of nine cases, and for two of the remaining issues, "conditions of contract" and "payment terms", this is most likely explained by the implementation of the "Housing Grants, Construction and Regeneration Act, 1996", which introduced statutory procedures for payment, contract conditions and adjudication in the United Kingdom. That this is a significant factor for the increasing compliance with the code in respect of contract conditions and payment terms is supported by the remainder of the survey indicators, which revealed general non-compliance with the code at a consistent rate[503].

The results suggest that, despite contractors' professed interest in closer buyer – supplier relationships, these remain traditional, arms-length and cost driven from the

---

[501] Ibid.
[502] Greenwood, 2001.
[503] Ibid.

outset. In the case of subcontract procurement, it may be wrong to dismiss all reports of new relational approaches and attitudes as "mere ceremony"[504], nonetheless it is consistent with the survey results above that a few contractors are experimenting with subcontractor partnering, while for the majority it is business as usual[505].

What has been reported of the nature of the main contractor – subcontractor relationship in the United States and United Kingdom can only be repeated when describing the situation in Germany, where a survey into the same relationship function has received the following responses from subcontractors[506]:

statement	experienced by number of subcontractors in %
prices being squeezed	88
pressurised for time to completion	75
poor site management by main contractor	70
poor treatment received	65
subcontract poorly defined	52
satisfactory co-operation with main contractor	50
main contractor was unreliable	47
treated as partners	37
main contractor became insolvent	30
main contractor kept to schedule/programme	28
payments on time	27

Table 11: Ranking of statements received by subcontractors

The relationship is rarely satisfactory, where only half of the subcontractors reported that they were co-operating satisfactorily with the main contractor and only approximately a quarter could claim that they received their payments on time or the main contractor managed to keep to the time schedule. Both parties engaged in legal as well as questionable practices in order to obtain an advantageous position and it is frequently the subcontractor who is the weaker party. This is to the detriment of both

---

[504] Institutionalists agree that a certain amount of organisational behaviour is really aimed at signalling legitimacy to key observers and confirming that the organisation conforms to expectations of how it should look and behave. Some aspects of organisational behaviour can be taken at face value, whereas others may represent a „largely ceremonial based structure" and involves simultaneous strategies of efficiency and leverage. Ibid.
[505] Ibid.
[506] Helmus and Weber, 2000.

parties, where the subcontractor being squeezed, the main contractor will suffer from poor workmanship and unsatisfactory service. Good site management and general capability on the part of the main contractor are very important aspects for efficient co-operation on site between subcontractor and main contractor. This is often influenced not so much by the particular firm but more by the individual site manager of that firm. Poor site management was identified elsewhere[507] as a characteristic that particularly weakened main contractor – subcontractor relationships. By contrast, good site management has a particularly positive impact on the subcontractors' ability to carry out their work. This requires adequate information to be made available in good time and competent main contractor representatives on site to co-ordinate and integrate all contributors effectively.

Reasons cited by subcontractors for experiencing problems were[508]:

response	given by number of subcontractors in %
financial structure of subcontractor	78
general business climate	72
dependency on main contractor	82
poor legal knowledge (of contractual rights)	63
foreign competitors	50
other	13

Table 12: Reasons for subcontractors' difficulties

This confirms that the financial situation is a matter of concern for the majority of subcontractors. Payment discipline, as shown by all surveys into the nature of the main contractor – subcontractor relationship, is extremely poor, which is of considerable concern to many subcontractors, who suffer from insufficient liquidity and are dependent on a regular cash-flow in order to manage flexible income and fixed payments[509].

---

[507] Kumaraswamy and Matthews, 2000.
[508] Helmus and Weber, 2000.
[509] Ibid.

Subcontractors have stated that they are not willing to work for some main contractors. The overall quality of co-operation is very much dependent on the main contractors' personnel involved in a particular project (including site manager, contracts manager, etc.). Factors that influence the willingness of subcontractors to bid on a main contractors invitation in order of importance are[510]:

factors	given by number of subcontractors in %
previous experience working with main contractor	90
resource availability	82
expected profit	78
schedule/programme criteria	77
site manager responsible for project	45
other	12

Table 13: Factors that influence subcontractors' willingness to bid

The evidence presented here offers a view that is partly contradictory in character on the nature of main contractor – subcontractor relationships, where on the one hand there is widespread and continuous use of subcontractors globally with some long-term business relationships occurring and on the other hand an adversarial attitude prevailing. The question whether a main contractor is to consider the price differential that he is prepared to pay for retaining the services of a trusted and capable subcontractor instead of choosing any lowest bidder, will be addressed in the following chapter. There are arguments for both alternative governance structures and it is more of a gradual change from one to the other rather than a clear cut decision, where a main contractor is faced with a continuous spectrum of business relationships when selecting subcontractors.

---

[510] Ibid.

6   Working with Subcontractors

6.1   Issues to consider when working with subcontractors

As was pointed out in chapter two, the Design and Build contractor, in order to deliver an integrated service, be the first point of contact for a client in need of construction services and who may be expected to extend the services offered to include the full support over the life-cycle of the building, has to concentrate his efforts on integrating the full supply chain, including design, technological expertise, management skills and business acumen. He has to optimise the use of preferred modalities of co-operation and be expert in handling subcontractors and suppliers as not only befits a single, but a succession of projects for a variety of clients. However, as the previous chapter illustrated, the majority of main contractors does not as yet seem to have grasped this precept to active competitive advantage. The wider appearance of contractor-led contracting, which stresses the integration of all contributors to a project, may bring about a change in behaviour, facilitated by greater potential and incentive for a main contractor to adopt a more intelligent procurement style.

6.1.1   The need for subcontracting

The reasons for subcontracting, the development of and the trend in favour of subcontracting across most construction markets are described at various points during this discourse where appropriate[511] and its need shall now be briefly discussed.

There are generally two significant barriers to the integration of production or in-house provision of services. One barrier is the limited access to capital, which especially forces construction organisations, who generally suffer from poor capitalisation and low levels of financial assets, to concentrate only on significant and strategically important functions of development and performance, since the integration of other, lesser activities would unnecessarily tie up capital and prevent the development of core competencies. The other barrier concerns the vital factor of flexibility of location for construction organisations, since a predominantly location based performance of construction services prohibits the complete provision of services with internal resources in all locations concurrently, especially when considering the cyclical demand

---

[511] See sections 2.2.3 and 5.3.2.

for construction. The following criteria affect the drive to either subcontract or not[512], bearing in mind, of course, all other restrictions facing a main contractor:

- The greater the specificity of an item, the closer integration will be.
- The greater the strategic importance, the closer integration will be.
- The greater the uncertainty is respect of qualitative, quantitative, timing or technical issues, the more difficult outsourcing becomes and the greater the benefits of in-house performance.
- The greater the frequency or regularity of use, the higher the tendency to integrate the performance, particularly in respect of specific and strategically important items.

Therefore, flexible resources on site are particularly important since classical[513] construction output cannot be produced in advance and stored for later consumption and a contractor has little control over size, timing and location of construction orders, other than accepting or declining the opportunity. The rate of utilisation of directly owned plant in construction, for example, averages about 60 % in Germany, which compares to an average rate of approximately 90 % in manufacturing sectors[514].

6.1.2   Risk management in procurement

There is no doubt that a systematic approach to risk management is a pre-requisite for effective procurement management. Risk management is used to establish project priorities, the roles of the various parties in the process and the number and type of work packages to achieve these aims. Thus, the key issues concerning procurement strategy can be addressed as follows[515]:

- division of responsibility,
- terms of payment,
- basis for subcontractor selection,

---

[512] also refer to: Picot, 1991.
[513] Classical construction here refers to site based construction operations excluding pre-fabricated components, which is, however, on the increase to overcome just this problem. The key to successful pre-fabrication is flexibility of components and fabrication methods, to ensure widest possible application and acceptance without excessive amendments.
[514] Herdt, 2000.
[515] Cox and Townsend, 1998, p. 227.

- degree of main contractor control / investment, and
- the most appropriate allocation of risk.

The construction industry has a history of frequent and excessive cost over-runs due to poor contingency management. Rather than the provision of meaningless percentage numbers, which are mainly based on management's perception of project risk, a less subjective approach to contingency allocation is called for.

Risks must be prioritised in an attempt to direct management efforts to those risks that may be effectively and economically managed. Each risk must be categorised in terms of its probability of occurring and magnitude of impact. Risks with low probability and low impact are effectively ignored as it is not considered cost-effective to manage them.

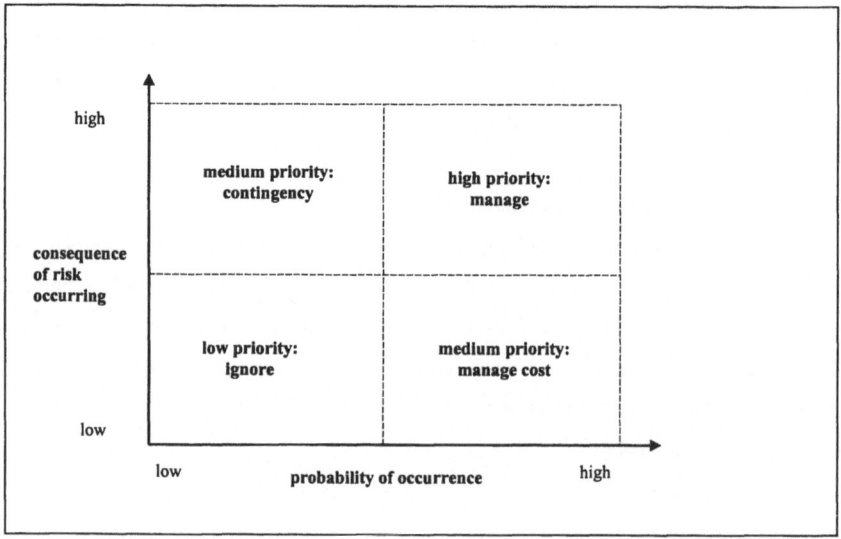

Figure 16: Prioritisation and management of risks[516]

The practice of risk management should be clearly linked with the complementary discipline of cost management and design/value management within the procurement

---

[516] Ibid. p. 233.

process[517]. The way that project losses due to badly managed risks are distributed confirms such an approach, where Holzmann in Germany, for example, claimed that 41% of its losses were down to estimating errors, 22 % of losses due to contractual risks, 30 % of losses were accounted for by poor construction performance and only 7% were caused by unavoidable acts of god[518]. Typically a number of questions will be raised:

- Is the base cost reasonable ?
- Can it be reduced by design?
- What are the risks involved?
- Can they be reduced / eliminated by design?

Thus, there are four linkages between risk management and the procurement process, which exist between risk management and the following three main areas of cost management, design management and contract management. Their interrelationship is illustrated below:

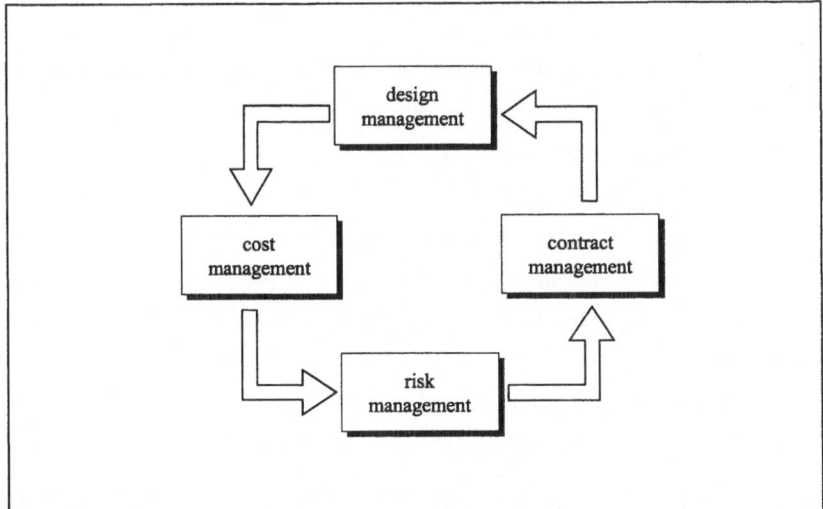

Figure 17: Interrelationship between risk, cost, design and contract management[519]

---

[517] see also 6.3..4
[518] Linden, 1999.
[519] also refer to: Cox and Townsend, 1998, p. 228.

This shows the interdependent relationship between risk, cost, design and contract management / strategy. The starting point is with preliminary design to an outline budget. This is costed and value managed to fit within the initial budget, providing the "true" construction cost. The process of risk analysis leads to the calculation of risk contingencies and the development of the contract strategy. The results of this exercise feed back into the design process and, by further value engineering, to bring total cost back into line with budgetary requirements[520].

6.1.3 The ideal and limitations of early supplier[521] involvement

The greatest benefit from early supplier involvement is obtained before construction starts on site. Where sophisticated clients bring the Design and Build contractor and his supply chain on-board as soon as the scope of the project is defined, they will be in a position to work together with the client and his eventual professional advisers to fully develop the functional brief, conceptual solution and the cost plan. The most important principle of early involvement is avoiding uncertainty, which is often the cause for cost overruns[522].

A Design and Build contractor to deliver effective cost management and improved functionality, resulting in a better value construction project, has to ensure that[523]:

- the functional brief for the project is accurately defined,
- account is taken of the through-life costs of the building project,
- cost-effective solutions to functional and technical requirements are provided,
- the participation in the project of those individuals and organisations who can demonstrate the necessary commitment and ability to meet the project's objectives is secured,
- sufficient financial resources are available,
- the contract programme is realistic,

---

[520] Ibid. p. 228.
[521] The term "supplier" is used in its broadest sense here, as it can refer to any form of contractor, consultant, subcontractor or materials and plant supplier at any position within the supply chain.
[522] see also section 4.5.3.
[523] Hill, 2000.

- sufficient time is allocated to planning the project before site work starts,
- the flow of information between the parties is prompt and accurate, and
- the interrelationship between the participants are understood and competently managed.

To ensure that the client receives value for money and that the project cost falls within the agreed budget, cost planning needs to become more sophisticated and is a service that the client is expecting from the contractor. To undertake such a service effectively it is necessary for the Design and Build contractor to be appointed at the earliest possible stage, in order to make a positive contribution at the brief and feasibility stage. The figure in 2.2.1 illustrates diagrammatically how the opportunities for making cost adjustments reduce substantially as the project progresses from the feasibility stage through to the end of the construction period.

As clients often turn to Design and Build as a means not only to obtain value for money but also to compress construction time and transfer risk to the construction industry professionals, earlier and more frequent subcontractor and supplier involvement is necessary. For most main contractors, subcontractors actually control the ability of the contractor to compete successfully on price, where subcontractors complete the majority, if not all, of the site construction. Design and Build requires the main contractor to select major specialist trade contractors for the planning and design phases well before price competition is usually possible[524].

Whereas a general contractor under a lump sum traditional contract has a contractual obligation and a central role in communications and co-ordination on site, it is a role based upon the traditional ideal of a fully documented design, which is rarely the case nowadays. Since much of the detailed design work and engineering drawings must be provided by the successful specialist trade contractor, their efforts can only be called upon after the contract is let under traditional procurement. This means that, regardless of the ostensible procurement system, with the exception of Design and Build, the contractor's responsibilities for issues of materials and workmanship and for the co-

---

[524] Kubal, Miller and Worth, 2000, pp. 297-398.

ordination of the specialist contractor's works becomes blurred with the design team's responsibility for issues of design and the co-ordination of the design process[525].

It is in the contractor's best interest to maintain progress and this requires involvement in the specialists contractors' design progress and an active part in communication and decisions between the designers and the specialist contractors. Where a contractor has little technical knowledge of the various specialist contractors' work and acts, in effect, as a post box, it is difficult and time consuming to achieve the required quality of both the specialist design and installation. The problems of split design during the construction stage are exacerbated where different specialist contractors' inputs need integrating. Diverse contributions need some mechanism for enabling mutual adjustment if their integration is to be effective. This is an organisational issue as it involves careful consideration of the way in which a project is split into specialist packages, as well as the timing of each input[526]. A Design and Build organisation with a sole responsibility for design and construction represents a suitable and appropriate platform with a focus on overall performance and due consideration to the organisational complexity of the overall process, by taking into account all design inputs, manufacturing schedules, delivery limitations, handling needs and assembly processes and increasingly extending to maintenance and operations as well. Designers and engineers working for the Design and Build contractor will propose interim designs for revision by the owner during the design development stage, in order to ensure that the design is developing according to his or her needs. At each stage of the design's development the contractor must review the changes for development-estimate compliance after which the design will be presented to the owner. The contractor should review each drawing with the client explaining each one in detail and projecting what the next phase of design development will reveal. All participating specialist contractors should be part of the contractor's review before presentation to the client, who must perform the same ritual of development-estimate compliance. Detailed minutes of design review meetings must be prepared and exceptions taken to the design development must be noted so that if further design development fails to incorporate

---

[525] Hughes, Gray and Murdoch, 1997, p. 45.
[526] Ibid. p. 45.

these comments, a record of having voiced concerns or even objection will be documented[527].

It is not unusual during construction for a client to request changes to the agreed design and / or specification. The Design and Build contractor must determine whether they fall within the context of fulfilling the client's intent[528] or whether these changes truly represent increases in the scope of work. There is often a very fine line between these two situations. When such client requirements occur, all specialist contractors affected should meet with the main contractor team to determine whether "they should have known" to incorporate this work, whereby all such changes must be made to comply with the intent of the client's requirements at no cost to the client. Conversely, if it is apparent that the requested change clearly exceeds the intent and scope of the Design and Build agreement, then two approaches can be pursued; the Design and Build team can propose a variation order for submission to the client, or can offer alternatives to the contract documents which would offset the additional costs of the client's variation request at no appreciable reduction in the quality of the project. If any changes are not handled in this manner as soon as they occur, the Design and Build contractor will find itself in a quagmire once the actual construction works are under way[529].

Because specialist trade contractors have focused on particular niche markets, they are often aware of new processes, materials and equipment that can improve the success of any project. Subcontractors participating in the planning phase can provide input into both procedures and systems that shorten the construction programme and can discuss how this work will be affected by other subcontractors so that potential delays can be identified before they harm a project's progress[530].

Virtual construction techniques including digital communication and mutual scheduling / programming also requires early and close relationships between main contractor and

---

[527] Levey, 1999, pp. 247, 266.
[528] The intent of the contract documents is to include all of the work required to complete the project, except as specifically excluded. No adjustment shall be made if the contract sum of a contractor has not been aware or anticipated work as may be required to provide a fully functioning building.
[529] Ibid. p. 248.
[530] Kubal, Miller and Worth, 2000, p. 338.

subcontractor. A Design and Build contractor cannot fully implement the hardware and software improvements possible without being connected electronically to all major subcontractors.

Subcontractors can provide information about competitors to main contractors that is helpful in determining the type of clients and projects competitors are targeting successfully. Subcontractors are often aware of strategic alliances being formed by other general contractors, new services competitors are providing clients with and even the financial condition of a competitor[531]. Subcontractors have in-depth market information regarding their specific niche within the construction community. They are interacting directly with various contractors, designers and clients. They are reading a diverse influx of information channels a contractor may not have access to. This different perspective of supply and demand of their sub-sector can shed new light on trends developing.

Specialist trade contractors are more directly involved with architectural and engineering organisations for Design and Build or Construction Management projects and alliances. Subcontractors can provide information about design teams they were working with successfully. Several specialist trade contractors have established themselves as the industry leader in their respective sector. Before any pencil is laid to paper, designers will call these companies first to collaborate their ideas and visions. Such a business partner for the Design and Build contractor can be invaluable since the partner is a key to unlock new opportunities and is preferred by designers by way of its experiences and successful track record[532].

The following schedule summarises the potential benefits that an early involvement of the supply chain in the design and construction process can bring[533]:
- increased certainty of out-turn cost,
- improved functionality,
- obtaining the most cost-effective solution,

---

[531] Ibid. p. 400.
[532] Ibid. p. 401.
[533] Hill, 2000.

- improved project delivery,
- improved quality,
- predictable through-life maintenance, and
- meeting or exceeding the client's expectations.

Further advantages that may benefit the supply chain include:
- greater certainty of repeat work,
- payment for pre-contract work,
- reward for good performance,
- improved margins,
- improved efficiency,
- reduction in waste of all kinds,
- non-adversarial supply chain relationships, and
- satisfied clients.

But in order to receive the kind of benefits just identified by way of early involvement of the supply chain, the Design and Build contractor has to be prepared to undertake the following[534]:
- Work closely with the client and possible advisors to understand, develop and deliver to the client's needs.
- Be prepared to share the benefits as well as the risks of collaborative working with the supply chain.
- Develop long-term relationships with his strategically important suppliers rather than selecting for the duration of one project only on the basis of "lowest price wins".
- Enter into mutually beneficial arrangements with fewer suppliers.
- A commitment to work with suppliers to improve value in project delivery.
- Assume responsibility for educating the supply chain in the techniques and changes necessary.
- Develop long-term strategic goals with strategic suppliers.

---

[534] Ibid.

While the case for co-operation and partnerships is put forward as a means to reap the rewards of early involvement in the development of a project and some other benefits through long-term strategic collaboration is not contentious, there are certain limitations to be considered in that partnership formations and maintenance are a costly process, require behavioural and procedural modifications on the part of both parties and are extremely difficult for small companies to develop, particularly where the suppliers are large organisations. Some business principles are universally applicable, such as the implementation of Total Quality Management, where the maintenance of high quality is an erstwhile goal of all parts of the organisation, but the same cannot be said of the function of long-term co-operative supplier relations.

There remains the tasks of finding suppliers / subcontractors that are willing to engage in partnerships, or at least strategic co-operations, where the purchaser's wish alone may not be enough. It is reasonable to assume that in order to justify a supplier's investment in the relationship, a buyer will usually need to be able to offer significant potential or actual filled order-books or profit. It is not enough for the buying organisation or its associated purchase expenditure to be large in an absolute sense. More generally, what matters is relative, not absolute, size. This can be measured in terms of the ratio of the main contractor's contract value placed with the subcontractor's total turnover figure[535]. No such data is publicly available in order to identify how large the figure must be before it has a significant effect on a supplier's response to the formation of a strategic coalition. It is safe to assume, however, that large global contractors will, in many of their supplier relations, enjoy a suitably large ratio of contracts or orders to be able to enter into strategic alliances including partnerships.

It appears, that if main contractors want to establish strategic alliances, they will need to offer a potential partner a significant ratio of value of contracts or orders to total turnover, or one or more of the following relationships characterise the supply relationship[536]:

- services are complex and involve a high degree of uncertainty,

---

[535] Ramsey, 1996 a).
[536] Ibid.

- where a stream of benefits is produced and accrue over time,
- where buyers seek to avoid significant transaction costs associated with multiple service ordering,
- where the market environment is turbulent,

to persuade the supplier that it is in its own best interest to invest in a long-term strategic coalition. In the absence of such pre-conditions it is predictable that suppliers will rebuff main contractors' co-operative advances.

Particularly larger suppliers are busy developing collaborative links with their major customers and suppliers and do not want to spend time and effort on developing similar relationships with smaller companies if strategically of no importance. It is, therefore, extremely difficult for small companies to develop partnerships with their suppliers, particularly where the supplier is a large organisation[537].

6.2 Selecting the right governance structure for main contractor-supplier business relationships

6.2.1 Current approaches

Traditional approaches to procurement on the one hand are essentially variations on a theme, where they assume that suppliers for each project are procured on an individual basis. The vast majority of construction work currently undertaken is procured in this "one-off" manner, with each party trying to extract a maximum reward for minimum risk. Main contractors currently using traditional approaches to procurement effectively apply a sourcing strategy that may be described as "adversarial leverage"[538]. Such arms-length supply relationships are usually only suited to non-strategic, low value and infrequent purchases, where there is a great deal of choice from a market of expert and capable suppliers, resembling a commodity spend. On the other hand, partnering relationships are frequently presented in purchasing literature and elsewhere as a generally applicable, universally desirable solution to the problem of sourcing strategy decisions[539].

---

[537] Ibid.
[538] Cox and Townsend, 1998, p. 40.
[539] Ramsey, 1996 a).

The problem is that the construction industry does not seem to understand that the correct way to think about procurement is to recognise that there is always a range of alternative procurement relationships available to deliver a particular material, plant, work package, design service or construction project and that it is not appropriate to assume that only one approach is always more appropriate than any other under all circumstances. This means that "partnering" may be an extremely valuable way of managing construction procurement under some specific circumstances, but it may not be under others.

6.2.2   Procurement classification and strategies for a Design and Build contractor

Analysis of procurement literature suggests, that there are a number of factors that determine the classification of a procurement transaction, which are identical to those that affect the decision to either outsource or integrate in the first instance[540] [541]. Combining essentially two models of procurement strategy models[542], the three dimensional matrix shown overleaf provides a classification of seven procurement strategy types, each a result of a combination of three basic characteristics.

The *strategic importance* of a given supply item can be defined in terms of the percentage of total purchase cost, impact on product quality[543] and business growth. The greater the significance of any one of these factors the higher the strategic importance of the supply item becomes for the buyer concerned[544].

---

[540] Cox and Townsend, 1998; Hamm, 1997; Picot, 1991; Pisoni, 2001; Ramsey, 1996 a).
[541] see also section 6.1.1.
[542] Cox and Townsend, 1998; Hamm, 1997; Ramsey, 1996 a).
[543] Product quality in this context is to be understood in terms of functionality and fitness for purpose meeting or exceeding the expected or specified standard.
[544] This is perhaps a worthwhile place to point out that although any strategically important factor is specific to a particular organisation, not every conceivable specific task is equally at the same time strategically important. Therefore, a degree of specificity in itself is not necessarily an indicator of a significant level of strategic importance to a firm, see also: Picot, 1991.

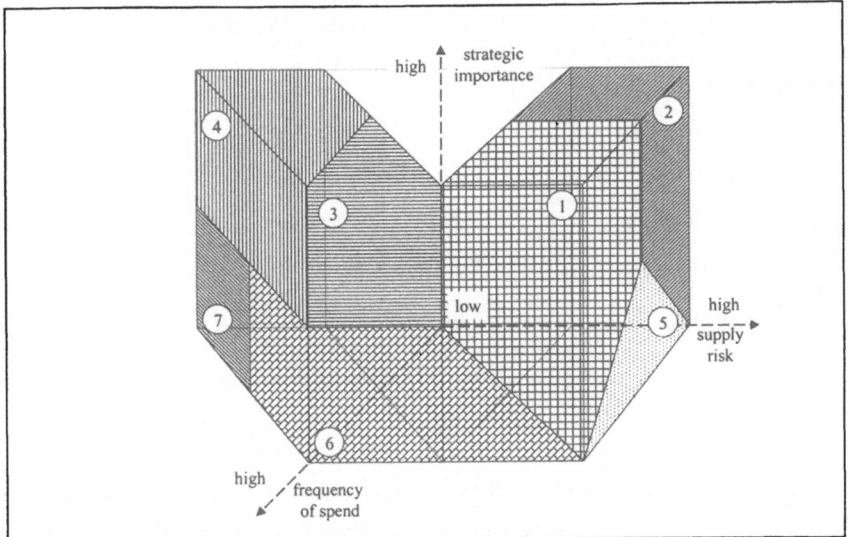

Figure 18: Procurement classification

The *supply risk* or potential market difficulty is addressed in terms of availability, number of suppliers, competitive demand, storage risk and substitution possibilities, which are all factors that characterise the choice in a supply market.

The concept of *frequency of spend* characterises the possible range of the type of procurement spending which is either one-off, occasional or a regular type of spend on a given supply item. It is not only an indicator for the internal procurement decision of an organisation, but also very much reflects the outward attractiveness of an organisations procurement choice. In other words, how well it will be received in the market or how much of an impact it will have in the market[545].

With the matrix it is possible to begin to differentiate between various procurement alternatives in relation to *the strategic importance, the supply risk and frequency of spend* as presented in the table on the following page[546]:

---

[545] see also section 6.1.3.
[546] Cox and Townsend, 1998; Hamm, 1997; Pisoni, 2001; Ramsey, 1996 a); Schulze, 1997.

	procurement choice	supply risk	strategic importance	frequency of spend
(1)	relational-led procurement	high	high	high
(2)	strategic procurement	high	high	low
(3)	preferred procurement	low	high	high
(4)	market-orientated procurement	low	high	low
(5)	bottleneck procurement	high	low	low
(6)	supplier-led procurement	low	low	high
(7)	non-critical procurement	low	low	low

Table 14: Procurement choices

For areas of regular spend (high frequency of spend), where suppliers are few in number and strategically important services are required, there is a need for close, long-term single sourcing or partnering. The buyer, on account of his large and regular spend, is in a position to attract the appropriate suppliers and enter to mutually beneficial arrangements, which can be described as *relational-led* procurement[547].

Where a low frequency or occasional type of spend prevails, or an organisation's impact on the market is very limited, it is unlikely to be a suitably attractive partner to enter into long-term partnerships on competitive terms. The organisation will have to seek a *strategic* procurement approach, which involves the identification of potential suppliers with whom closer co-operation on suitable projects may be possible for mutual benefit and can also be referred to as dependent sourcing. An intelligent selection is required of suitable suppliers with whom closer co-operation should provide a means of accessing their technical expertise and design knowledge. A relationship resembling a strategic alliance is called for, which requires a degree of market knowledge and market research[548] to identify the leading suppliers at any point in time, demanding intensive communication and an early exchange of ideas and information.

---

[547] see also section 5.3.1.
[548] see also section 6.2.4.

For areas of regular spend, where there are potentially many proficient suppliers to choose from, there is a possibility for longer-term supply relationships in the form of *preferred* suppliers. This relationship can also be referred to as leveraged purchasing, since the bargaining position of an organisation with sizeable and frequent type of spend can negotiate favourable terms in exchange of certainty for its suppliers[549]. An important concept to distinguish this style of procurement to that of supplier-led procurement is that the quality of the suppliers' products or services needs to be consistently of the highest order.

A *market-orientated* approach to procurement is appropriate in circumstances where suppliers of significant strategic importance, principally in terms of value and time constraints, are only occasionally called upon and the buyer experiences only low levels of supply risk. There are a number of competing suppliers in the market place who supply a similar quality product or service and procurement occurs on short notice on the basis of standard specifications and best value for money. An arms-length, multiple source supply relationship is sought and the purchaser continuously seeks to add to the number of competing suppliers.

The distinguishing characteristic of *bottleneck* purchases is that although they may not represent a strategically important supply item generally, they still constitute a serious risk to the organisation in that it is difficult to obtain and thus poses a high degree of uncertainty.

A relatively close, preferred *supplier-led* relationship is appropriate for services and products that are frequently required, however, are of low strategic importance and widely available and thus should involve the lowest possible transaction costs. Therefore, a supplier-led procurement relationship should be carefully co-ordinated and ensure a seamless supply of products and services on preferential terms to the purchaser. Such a position is possible if the buyer represents a sizeable proportion of the suppliers turnover.

---

[549] see also section 5.1.2.

Non-critical procurement requires an efficient sourcing approach of supply items where costs of procurement is often greater than the value of the product itself (e.g. office consumables, sundry supplies to site, etc). The aim here is to reduce transaction costs, possibly with a framework agreement or transfer of procurement to internal customers.

One of the difficulties is trying to determine what proportion of a company's purchases and orders are likely to be suited for a particular approach. For example, closer supplier relations are associated with uncertainty, significant strategic importance or a frequent type of spend, but it is not known as to the amount of any of these factors that is needed to trigger the move towards closer relations. There exists, therefore, a problem of calibrating the operational dimension of these concepts. Reference to conventional ABC analysis[550] may be of use and would indicate that a relatively small proportion of the number of supply items purchased by any organisation will normally account for a large proportion of the final outlay. It could be argued that items falling into the top 20 % to 30 % of an ABC analysis may normally be described as being members of the class of purchases deserving the title high value product or percentage of total purchase cost. As a result, *relational-led* and *strategic procurement* are likely to predominantly belong to this group.

6.2.3 A Design and Build contractor's procurement choices

The classification on the following page is an example of procurement strategy choices for a Design and Build contractor concentrating on two specialist trades, structural steel work and facade, which is a typical scenario for a commercial office development.

It is generally the case that construction organisations experience considerable fluctuations in their frequency of spend on suppliers from an organisation's perspective, if not from a project view. This is one of the very reasons for subcontracting and outsourcing to occur in the first place as described previously. The only exception are sundry supplies of low supply risk and strategic importance, which apply to a

---

[550] An ABC analysis is a hierarchical ranking for the selection of products or services, which are to be analysed for added value. Three categories of A, B and C are formed, which satisfy the most important, important and unimportant items. The following selection criteria may be used for example: turnover, overhead contribution, cost content or value to quality relationships. Brüssel, 1995, p. 1.

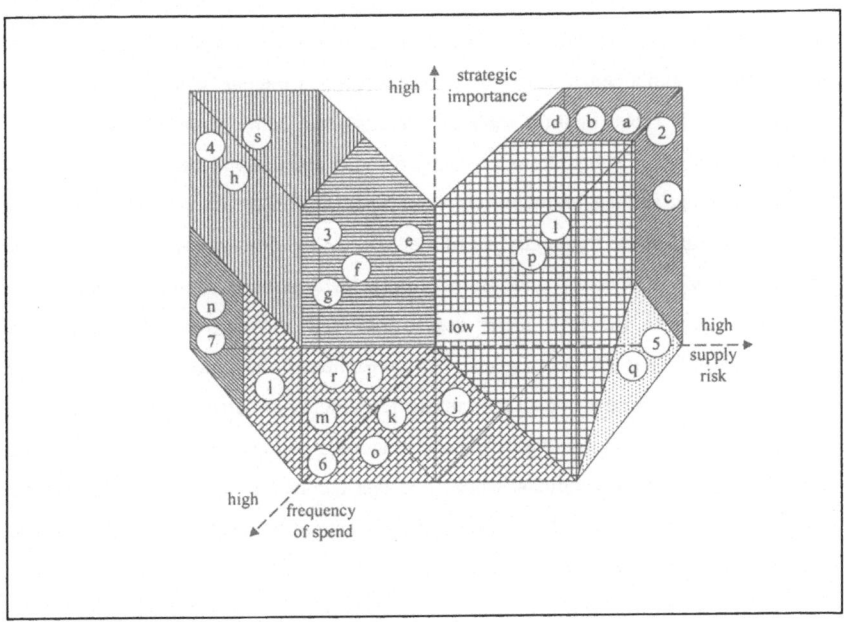

Figure 19: Example of procurement strategy choices

	subcontractor/supplier	supply risk	strategic importance	frequency of spend
a)	architectural and design services	high	high	low
b)	mechanical and electrical services	high	high	low
c)	facade (metal, glazing, natural stone)	high	high	low
d)	structural steel	high	high	low
e)	ready-mix concrete }	low	high	high
f)	reinforcement }[551]	low	high	high
g)	cement }	low	high	high
h)	fit-out trades	low	high	low
i)	site set-up	low	low	high
j)	energy	low	low	high
k)	company vehicles	low	low	high
m)	office supplies	low	low	high
n)	site timber	low	low	low
o)	management travel	low	low	high
p)	information technology	high	high	high
q)	disposal	high	low	low
r)	logistics	low	high	high
s)	plant/equipment	low	high	low

Table 15: Main contractor's subcontractors and suppliers[552]

---

[551] Ready-mix concrete, reinforcement and cement are often performed as a single package by a framework subcontractor.

[552] adapted from: Pisoni, 2001.

contractor's operations across all projects. In respect of key strategic suppliers, for reasons already explained, a contractor is faced with an irregular type of spend. As such the procurement classification above demonstrates how *strategic procurement* dominates, that other subcontractors, materials and plant suppliers are split between *market-orientated* and *preferred procurement* types relative to frequency of spend, and items of low importance frequently are of the *supplier-led procurement* type. This is not surprising, since a Design and Build contractor's principal procurement is for subcontract services and to a lesser degree materials and represent from a low of approximately 60 % to 70 % in Germany to approximately 80 % to 90 % in the United States and United Kingdom of all procurement, including architectural and engineering services. It is typical for builders plant and materials to represent the bulk of market orientated and preferred products, while specialist contractors such as mechanical and ventilation, facade and structural frame subcontractors are usually counted amongst strategic procurement services.

6.2.4 Systematic approaches to procurement market research

The matrix shown over the page serves as a suitable summary to illustrate the preferred marketing approach to procurement[553].

The matrix illustrates in a fairly self explanatory manner how market research shall be basically performed, except to say that in the event of outsourcing all the strategically important and low supply risk items, the chances of success to come to favourable terms will very much depend on the relative frequency of spend or market impact. Where the company is in the market for only irregular or occasional purchases, it will be in a better position to undertake these itself at lowest possible cost, rather than spend a prohibitively high price for an outsourcing service.

---

[553] Pisoni, 2001, p. 31

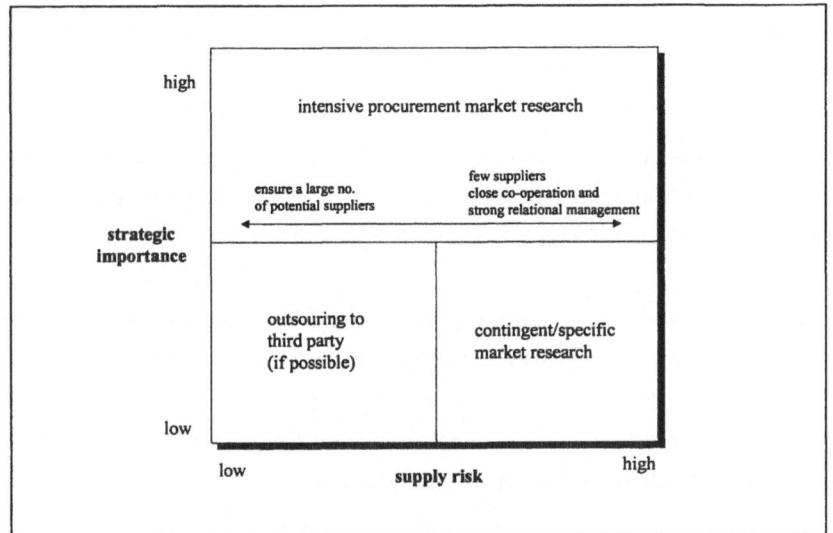

Figure 20: Preferred marketing approach to procurement

6.3   Behaviour and control exercised in contractor-led procurement

A Design and Build contractor, aware of the risks involved in being the project's sole supplier of design, construction and possibly maintenance and operation as well, needs to be very cautious in the selection of projects and clients and at the same time has to satisfy himself that the project suits its particular skills, meets business objectives and has adequate management and resources available to support the project at the level required to satisfy the client. If these prerequisites are satisfied, the Design and Build contractor is very much concerned in assembling his project team and supply chain, which must include design and specialist contractors contributing to the overall design package.

6.3.1   Good tendering and estimating practice

A Design and Build contractor, whose obligation involves the supply of the design, must involve specialist contractors' design input, which often is a critical element in the construction process. Where such specialist design is required, it is desirable if not essential that specialists are involved at the earliest possible stage. Adequate time for tender and design preparation is critical and the design team must be provided with clear briefing.

It must be remembered that copyright of any design prepared by tenderers will automatically rest with them and will not transfer even if they have been paid for the design, unless such a transfer has been expressly agreed[554].

Good practice on the part of a client includes the provision of adequate time for each stage in the selection of appropriate suppliers. Clients should appreciate the close relationship between the time allowed for the preparation of tenders and their subsequent quality. Where a contractor has insufficient time to follow the appropriate procedures, the final quality of a project is likely to suffer.

The criteria to be used in assessing subcontractors' tenders shall be notified during the selection process and stated in the tender enquiry documents. These criteria for qualification should include: the quality of work, performance realised, overall competencies, health and safety record, financial stability, appropriate insurance cover, size and resources, technical and organisational ability and the ability to innovate. The process of qualification is important whether subcontractors, including architects and engineers, are to be selected to tender competitively or are appointed on any other basis, e.g. by negotiation[555].

Potential tenderers require sufficient information to enable them to decide if they want to tender, which should include[556]:
- job name and location,
- nature, scope and approximate value of the subcontract works including reference to the extent of any design work required,
- likely dates and duration of both the tendering process and the subcontract works,

---

[554] Construction Industry Board, 1997, p. 8.

[555] e.g. CIOB, 1997 a), p. 53; CIOB, 1997 b); Construction Industry Board, 1997, p. 9.

[556] Ibid.

- the number of tenderers invited to submit a formal tender[557]:

subcontract type	max. no. of invitations to issue	min. no. of compliant tenders required
design only	4	3
construction only (including some minor design/proprietary supply)	6	4
design and construction	3	2

Table 16: Number of tenders recommended

- whether the contractor is already appointed or is involved in a tendering process,

- main contract tender date,

- approximate value and period of the main contract,

- whether, and how, any costs may be shared,

- whether the tender will be based on a bill of quantities of other pricing documents or a specification and drawings, or specification only,

- selection procedure and selection criteria,

- main and subcontract conditions, and

- name of the client and relevant consultants.

Briefing sessions may be appropriate where they can provide additional clarity and information for either party and thereby increase the likelihood of compliant tenders. This is especially relevant where a project is large or complex, or where a specialist contractor will have substantial design input. If the parties involved are not familiar with each other, such sessions can also help to establish more clearly whether they would be suitable and compatible team members[558] and help to settle any outstanding questions in respect of the subcontract documentation[559]. A list of reserve tenderers should be prepared, one or two for each package or trade, and should be informed that they have been selected reserves and that they will not be asked to tender unless any of those on the tender list will drop out.

---

[557] Lists should be as short as possible, consistent with the objective of receiving a sufficient number of compliant tenders. They should generally be selective where the requirements are more complex and, therefore, the tendering process more costly.

[558] Ibid. p. 12.

[559] CIOB, 1995.

Proposed subcontracts should be compatible and consistent with the main contract and suits of contracts and standard unamended contract forms from recognised bodies should be used where they are available[560] [561].

The time required for tendering will vary according to the precise nature of the project. More time, however, may be required to prepare tenders where a project is large and/or complex, where specialist design is needed, or where products and materials have to be sourced from unfamiliar or distant suppliers or subcontractors. Suitable periods to allow for tendering for most projects are given in the table below[562]:

subcontract type	minimum tender times in weeks
design only	3
construction only (including minimum of design/proprietary supply)	6
design and construction	10

Table 17: Tendering times

Key principles of good practice that apply to tender assessment, particularly from a Design and Build contractors point of view, are that[563]:

- conditions for all tenderers should be the same,
- confidentiality should be respected by all parties,
- tender assessment should have regard to quality as well as price,
- practices that avoid or discourage collusion shall be followed, and
- tender prices should not change on unamended scope of works.

When the lowest tender received exceeds the client's budget, changes should be negotiated with the lowest tenderers. This process is either based on recommendations from the subcontractors for cost savings or design changes, which reduce the scope or specification of the works in advance of a firm price agreement between Design and Build contractor and client. Only if significant changes are proposed to a scheme, two

---

[560] Construction Industry Board, 1997, p. 15.
[561] see also chapter 3.6.
[562] Ibid. p. 17.
[563] Ibid. p. 19.

or at most three tenderers may be asked to re-tender in competition, thereby retaining the lowest market price for the amended project[564].

Estimating is defined as the technical process of predicting cost of design and construction[565], where the management is an important element in its production. Estimators must have management responsibility within the department or group responsible for estimating and for managing the production of the estimate, ensuring that other contributors work to their requirements, produce information on time and in the format required, so that effective operating procedures and lines of communication are established between all contributors to allow the efficient production of estimates[566].

A Design and Build contractor must prepare an estimate in a way that is explicit, consistent and takes account of design issues, methods of construction, through-life performance and circumstances which may affect the execution of the works on a project. A reliable project estimate can only be achieved where each operation or item is analysed into its simplest elements and the cost calculated methodically on the basis of factual information. Any other method may be suitable for arriving at an approximation of the project cost, suitable for setting an overall budget, target costs or other preliminary estimate, but are inherently unreliable and should be approached with caution.

The decision to tender should, therefore, not be one that is taken lightly, instead all contractors should have a strategy expressed in a corporate plan. Some contractors' corporate strategies are more detailed than others, but it should give details of a company's turnover target broken down into various divisions or sectors of work. Against this corporate plan senior managers will take the decision to bid for a specific contract based on the following factors[567]:

- the potential contribution of the contract to the company's turnover in a particular sector, the overhead recovery and the anticipated profit,

---

[564] CIOB, 1997 a), p. 183.
[565] Tendering is a separate and subsequent commercial and management function based upon the net cost estimate.
[566] Ibid. p. 1
[567] McCaffer and Baldwin, 1995; pp. 37.

- the likely demands of the contract on the company's financial resources,
- the company's available resources,
- the type of the work,
- the location,
- the client, and
- the contract details.

Contractors will want to avoid contracts that are too large for their size, beyond experience range, stretch available resources too far – including cash, are well outside their normal geographical area of operation and contracts that have unusually onerous conditions of contract[568].

A Design and Build contractor's strategy for the selection of bids should be one of fewer bids with greater reliability, rather than many unreliable bids. The following figure illustrates this well[569].

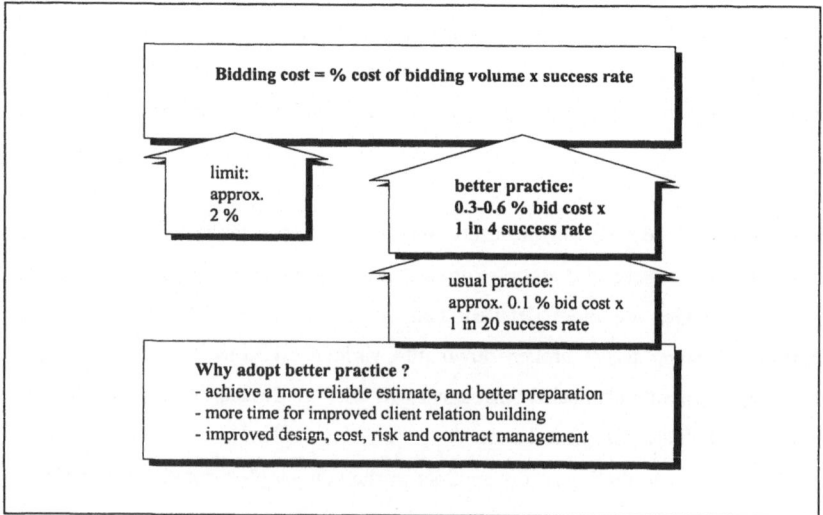

Figure 21: Relationship between bid frequency and bid cost

[568] CIOB, 1997 a), p. 32.
[569] Jacob, Winter and Stuhr, 2002, p. 22.

6.3.2 Competitive versus negotiated supplier selection

Market-orientated procurement is characterised by competitive, single-stage tendering, which is generally the most appropriate form of competitive tendering, but two-stage tendering may be suitable for larger and/or complex packages of greater significance to a project, where early involvement of the subcontractor is required prior to the completion of the full design, as is the case with preferred procurement. Where the early involvement of a specialist contractor is required for a vital design service or other significant specialist input, this is best provided for by direct negotiation between the parties on a separate selection process[570], as suits strategic procurement requirements and the occasional bottleneck purchase.

With two-stage tendering[571], the first stage of selection is based on pricing documents related to preliminary design information. Those provide the level of prices on which to base a final price once the design has been completed. Selection for the second stage does not imply that a contract for the works has been entered into[572].

Competitive tendering may be impossible or inappropriate, for example, where only one organisation has the expertise or resources required[573] or where products or services are required urgently and there is not enough time to undertake the competitive process properly[574].

6.3.3 Supplier appraisal and development

Supplier appraisal and development are not the same thing. Appraisal involves some form of assessment against a certain standard. Supplier development is the process where a partner in a relationship modifies or otherwise influences the behaviour of the other partner with a view to mutual benefit and involves the following activities[575]

---

[570] Construction Industry Board, 1997, p. 8.

[571] A method a client can use to introduce an element of competition into the selection of a Design and Build contractor to be solely responsible for whole project delivery (see also advantages and disadvantages 4.3.2 and 4.4.2 respectively).

[572] see also section 3.3.3.

[573] Such as relational-led, strategic or bottleneck types of procurement relationships.

[574] Ibid. p. 8.

[575] Cox and Townsend, 1998, p. 240.

- supplier co-ordination, moulding the entire supply chain into a common way of working, and
- individual supplier development, to help improve the strategy, tools and techniques used by a particular supplier.

A process of systematic valuation of supplier performance data enables the buying organisation to negotiate agreements covering required improvements in cost, time , quality and other performance criteria with longer-term or strategic suppliers. The benefits of this approach include[576]:

- on-going cost reductions,
- supplier innovation to improve product/process performance, and
- improvements in the system/processes of the buying organisation.

The technique of supplier development, however, is only suitable under certain circumstances as the full development programme is costly and only likely to yield results with relational and strategic supply relationships. Primarily, the approach requires a high degree of co-ordination and co-operation and the application of scarce and valuable human resources becomes necessary to achieve continuous improvements in the performance of buyer and supplier alike[577].

In recognition of the growing role of suppliers to the successful performance of any business, but especially to a Design and Build contractor, there is a need for an objective assessment of strategic suppliers and their performance in meeting the expectations of the client.

6.3.4   Early involvement tools

The circumstances that favour the implementation of early involvement tools have been referred earlier and shall now be briefly described.

---

[576] Ibid. p. 243.
[577] Ibid. p. 224.

Partnering

Involves two or more organisations working together to improve their performance through mutually agreed objectives, deciding on a method for resolving any disputes and commitment to continuous improvement and sharing gains. It is essentially about communication and can be extremely demanding and relies heavily on trust[578]. A very important aspect of the partnering approach is related to subcontractor identification, selection and appointment. This involves work package and company identification, where all major packages on the project that can benefit from the partnering approach need to be ascertained. Trade packages are examined under the headings of design content, complexity of construction, high subcontract value, long periods of construction, early commencement of construction, high levels of aesthetics, long procurement times and those trades that could add value with their early input. Key trades/packages that are usually identified include, for example, mechanical and electrical services, structural steel frame, brickwork/metal/glass facades, natural stone cladding, atrium glazing and lifts/escalators.

As a means to achieve the objectives set out in a partnering agreement one or more of the following techniques and methods are implemented, confirming the view that the process of "partnering" refers to a combination of individual business tools. The more of these business tools are actually adopted in a business relationship the greater the chances that it resembles a true partnership.

Value Management[579] (VM)

A proactive, creative problem solving process. A VM study aims to attain optimum value by providing the necessary functions at least cost, without prejudice to specified quality and performance. Value, thereby, is a concept based upon the relationships between satisfying needs and expectations and the resources needed to achieve these. The best results with VM are achieved where it is applied as early as possible in the project process and involves the supply chain[580].

---

[578] Hill, 2000.

[579] For further, more detailed information on the background and implementation of Value Management and Value Engineering, see: Male, et al., 1998 a); b).

[580] See also Figure 3 in section 2.2.1.

Risk Management

A process for identifying activities that may have a negative business impact and developing a strategy to minimise or eliminate the potential effects[581]. To actually identify all possible risks every contributor with significant input to the project has to be present as early as possible.

Whole-Life Performance

A means of comparing design and construction options with their trough-life costs and future performance. It promotes the selection of design and construction solutions that meet the performance requirements for the project at the appropriate correlation between investment and maintenance/replacement cost, which can mean the lowest through-life cost. It needs to be performed alongside design development from an early stage to be effective.

Continuous improvement

An umbrella term for a number of tools required to ensure that a task is executed better each time it is performed. Its aim is to identify problems before they happen, rather then after, and utilises the experience of the supply chain to continuously seek better ways of doing things. This requires their early involvement in the project development process.

Benchmarking

A Total Quality Management tool used to measure and compare an organisation's processes (business, managerial or operational) with those of other organisations to deliver better processes and improved strategies.

Key performance Indicators[582] (KPI)

One of the tools used in Benchmarking. By assessing performance based upon a set of key performance indicators clients, their professional advisers, contractors and suppliers can measure their own performance and that of their construction supply chain relative to others in order to identify areas where improvement is required. Typical KPIs

---

[581] See also section 6.1.2.
[582] For further information about KPIs, refer to: The KPI Working Group, 2000.

include: client satisfaction – product, client satisfaction – service, defects, predictability – cost, predictability – time, profitability, productivity, safety, construction – cost and construction – time.

6.3.5 Control of the project development process

Three key factors have been suggested that must be controlled for a successful project completion by a Design and Build contractor[583]:

- Detailed design programmes must be used to enable all aspects of the design to be completed and integrated on time. The usual problem for a client to ensure that the responsibility for preparing and monitoring the detailed design programme is entrusted to the appropriate person, is easily resolved when pursuing in a Design and Build project. He only has to co-operate with a single party, which is responsible for all preparation, performance and control, and, therefore, must have access to appropriate design resources.

- The interfaces between work packages must be adequately predicted and defined. This means that all the requirements of each specialist contractor can be fully documented and made available to the preceding trades at the beginning of their own work.

- Possibility of long term detrimental effects requires specialist study to ensure that compatibility of the physical properties of materials and components between work packages is maintained, without possible long term interaction that are deleterious to the finished building.

While the Design and Build contractor is ideally placed to manage the integration of specialist contractors and other suppliers with the design team, the task of managing design services is inherently difficult because the design process itself is non-sequential and interactive and the process is dictated by the specific needs of a project. A Design and Build contractor should be successful, if he heeds the following[584]:

- Implements good programme management, which requires control, motivation and intervention on the part of his project management team.

---

[583] Hughes, Gray and Murdoch, 1997, p. 45.
[584] Ibid. p. 52.

- Timing of the appointment of specialist contractors should be governed by the design sequence, not the construction sequence.
- A lack of understanding and control of the interfaces greatly increases the potential for failure in the long term of incompatible systems and/or materials and the fixings/installations thereof[585].

A complex design involving a sequence of assembly steps on site between which other specialist contractors' has to take place, has a high potential for discontinuity and consequential low productivity. The site assembly process can become a series of small and inefficient operations separated by idle time waiting for the next operative to be available. Such a waste of resources can be avoided by involving the specialist contractors early in the design process, who should ensure that buildability studies are developed in conjunction with the contractor's design team with the objective of keeping the site handling, assembly and fixing processes as simple as possible [586].

The flow of information must be co-ordinated to ensure the right information is available at the right time to allow the design team to make correct overall judgements. The timing of the exchange of specialists' technical and dimensional information is often critical to the completion of the whole design. The necessary project management of the flow of this data is usually outside the scope of a traditional general contractor's responsibilities. There are few examples of the complete design process being planned and managed as a single integrated process apart from Design and Build projects[587].

A contractor-led approach further prevents the problem of a frequent lack of clarity occurring in respect of the legal implications of design approvals and different opinions as to the level of checking that is necessary, particularly where responsibility of design is split between designers and specialist contractors, as the Design and Build contractor is solely responsible.

---

[585] For example, where the procurement management of the specialist contractors has failed to bring them into the development process at the right time for dimensioning to be clarified in the design process, physical clashes can easily arise on site. These lead to delay whilst they are resolved and site adaptation of prefabricated components may be necessary, negating the advantages of prefabrication.
[586] Ibid. p. 48
[587] Ibid. p. 49

Design issues, interfaces and checks must be jointly resolved between the contractor's design team and specialist contractors at the right time in the procurement process and with the right level of detail. For this to come about in an orderly and proper manner, two conditions must be met: first, the design team must specify what they themselves have done, and the result they consequently require from the specialist contractor to complete the design; and second, the specialist contractor must be sufficiently proficient at design and have the requisite technical knowledge and resources to respond and provide the necessary level of support to the design and subsequent execution on site. Ideally, designers prefer specialist contractors with whom they can work towards developing a solution jointly, and specialist contractors expect that, as a consequence of this, there will be a lower rejection rate of the developed designs they submit for final approval[588].

---

[588] Ibid. p. 51

7   Review

7.1   Summary

It is the action of forces of the environment on the client's organisation which is at the
root of the process of providing a project, whether it is a response in order to survive,
take an opportunity to expand or become more efficient, and will as a result require
construction work to be undertaken, providing the construction industry with work.
Whilst this process should be an open adaptive system to suit the client and the project
at hand, it is in practice always constrained by the environment within it exists, which
varies from one market to another. Construction markets are structured into construction
organisations of varying sizes and a series of project based vertical markets, where
contractors are highly fragmented at the lower end, but as project size and complexity
increase and geographical perspective widens they are more concentrated as
management experience and access to financial markets becomes critical. In addition to
fragmentation in size, the construction industry is made up of a number of participants
that not only include clients and contractors, but also consultants, including a variety of
architects, engineers, project managers, cost consultants (quantity surveyors), property
managers, and in addition material and plant suppliers and specialist contractors, who
often perform the role of subcontractor. The contribution of all of these participants
influences the process of providing a project.

Clients of construction services at large are generally not particularly satisfied with the
results of the construction industry in terms of either cost, time or quality. While it can
be agreed that many of the problems encountered by clients are down to their own
behaviour when procuring buildings, it is experienced clients who drive the stimulus for
innovation in construction procurement. Rather than turning to consultants for specialist
advice in every case, they are realising that aligning strategic and operational practice
with a portfolio of procurement systems points the best way forward to achieve a
desired corporate outcome.

A general procurement model for the selection of the appropriate procurement path has
been presented, which upon consideration of a number of variables to a set of eleven

client and project criteria, will identify the procurement path or choice of procurement paths that should be worthy of serious investigation.

At the upper end of the market construction consultants, managers and contractors have adopted over time a range of construction development and realisation methods, each with their particular strengths and weaknesses, which have been broadly classified into three groups of procurement types: designer-led, management-led and producer-led. The trend at the upper end of the market towards fewer but ever larger consultants and contractors, who aim to offer a total global solution to the construction needs of fewer, yet more powerful and increasingly demanding clients, has been described and illustrated with a number of examples. There will always be smaller, inexperienced and one-off clients, who can benefit from the choice that the development of a variety of procurement methods has brought about.

It is the producer-led approach from the range of available procurement paths, including Design and Build, Turn-key and BOT, which has been shown to be particularly suitable in promoting an integrated service. This scores highly on aspects of price certainty, timing, contractor input, management, risk avoidance, operation and maintenance (in the case of BOT), and when applied conscientiously on aspects of complexity, quality and competition as well. A degree of controllable variation is possible as long as the basic project parameters remain true to the initial brief, otherwise other aspects will be affected.

A construction process led by the producer, with responsibility not only for construction, but also for design and possibly for its performance as well, can provide the key to improve effective integration between client and the construction supply chain, since it offers a closer focus to all involved. Any contractor that has positioned itself in the lead role of construction procurement, as either a Design and Build, Turn-key or BOT contractor, must take note of the views of a variety of clients[589] and consider those that are to be targeted at all times. Obviously, clients vary in many ways,

---

[589] Broadly speaking, there is the choice between experienced and inexperienced private, corporate or public clients, with either frequent or occasional spending on construction services in a number of different market segments, locations, etc.

not only in terms of objectives that they seek to satisfy, but also in differences in their experience of the construction process, the importance of the project to their value system and whether they are one-off, casual or repeat clients with a high and regular construction spend. Very often, the eventual outcome of a project is determined by the worst performing partner and this includes the client.

The traditional separation of the process into design and construction in a project establishes an arena where control of the project is a potential source of conflict. Thus, the co-ordination of the integrative process between designer and contractor is seen as one of the major areas of difficulties, delays and disputes. A producer-led approach has been shown to be ideally suited to ensure that a comprehensive management approach is established at the outset to facilitate the proper integration of inputs from all contributors to the design and construction process.

The reasons for subcontracting to exist and its proliferation have been discussed and the need for specialist contractors of high calibre and with appropriate resources to execute the necessary works was identified. The characteristics of main contractor-subcontractor transactions of high asset specificity and uncertainty coupled with specific quality objectives, budget restrictions and time constraints present numerous challenges. A main contractor can address these challenges by establishing good business relationships with strategic subcontractors and suppliers, since relationships of high quality facilitate the function of subsequent transactions. Unfortunately, the current nature of main contractor-subcontractor relations was found to be still largely traditional, arms-length and cost driven from the outset, resulting in adversarial relationships, despite contractors' professed interest in closer buyer-supplier relationships.

Despite the recognition that specialist contractors, suppliers and designers are all necessary for the provision of design and construction services, their capability bearing directly on the quality of a project, traditional approaches to procurement still remain the standard in the majority of situations. There appears to be a problem in that the construction industry does not seem to understand that the correct way to think about procurement is to recognise that there is a range of alternative procurement relationships available to obtain a particular service or product and that it is not appropriate to assume that only one approach is always more appropriate than any other. For this reason a procurement classification model has been presented, which offers a number of strategies for a producer of construction in respect of the supply risk, strategic importance and frequency of spend on a particular service or item for a project or series of projects.

7.2   Conclusion

Construction is a saturated market nowadays, with the exception of some specialist services represented by proprietary process technologies or management expertise in delivering large and complex projects on a life-cycle basis. At the same time clients have been found to state that "the construction industry is too complex, costs too much money and does not deliver what it is expected to deliver", where clients are often confused by an increasing number of participants and each person in the construction team wanting authority over the project, but very few prepared to take financial responsibility.

With the change from a sellers' to a buyers' market and clients facing a greater choice than ever before, as the construction industry has become global and more complex at the upper end of a hierarchical, vertically structured market in response to clients organising construction work into fewer but larger contracts with more risk transfer and responsibilities, consultants and contractors alike have moved towards multi-disciplinary teams offering design and management services, challenging single service consultants and contractors. As a consequence subcontracting is on the increase and on the one hand medium sized organisations are disappearing, where a consolidation of larger firms absorb smaller ones, either to provide access to new geographic areas, new market segments or new clients, and on the other hand a specialisation into specific skills or geographical locations is occurring.

Acknowledging the fact that there is no homogenous market of either clients or construction service suppliers, nor a single best practice approach to the procurement of construction services, a general procurement selection model has been established to identify the appropriate approach to the procurement of a specific construction undertaking.

When considering all of the above, it appears that meeting clients' demands for a ready purchase of design, procurement and management of construction from a single source is most appropriately accomplished by the adoption of a producer-led procurement path, especially when expecting higher levels of efficiency, cost certainty, punctuality and quality levels. This can be achieved either through Design and Build, Turn-key or BOT depending on the preferences of the client and the needs of the project.

While better practices in choosing a producer-led approach to procurement are recommended, especially ensuring earliest involvement of an experienced contractor on the basis of a well thought-out functional brief (scope package), alternatives are possible and can sometimes even be desirable under specific circumstances, such as the need for accountability of a public client or third party advice to inexperienced clients. If, however, uncertainties exist as to the client's requirements and substantial variations to either scope or timing of execution of the project are expected, then it is unrealistic to expect the benefits from a producer-led approach to be forthcoming and a more reactive style of procurement is more suitable, as offered by management methods. In that case, however, without the cost and time certainty or convenience of a single source of responsibility for design, construction and possibly operation as well.

Finally, just as clients face a range of procurement options, main contractors have to find suitable suppliers, for whom it is essential to understand the process of determining the most appropriate types of relationships they require. A competent practitioner will need to know when it is safe to single source from a supplier, when it is appropriate to undertake joint ventures or when preferred suppliers or market-place supplier tendering is the most effective way of sourcing a construction project. He has to optimise the use of preferred modalities of co-operation and early involvement and be expert in handling specialist contractors and material suppliers as not only befits a single, but a succession of projects for a variety of clients and project types. The majority of main contractors, however, do not as yet seem to have grasped that an intelligent approach to supplier business relationships is a pre-condition to active strategic advantage. A first step into the right direction can be made with the help of a classification model, which has been presented to illustrate the appropriate use of procurement strategies in respect to parameters of supply risk, strategic importance and frequency of spend on a particular service or product.

7.3   Outlook

It is to be anticipated that the development trends identified and presented here as they refer to construction markets, industry and participants, will continue in the direction described, with some minor deviations in degree from one market to another. The gap between what have been described as experienced-frequent and inexperienced-occasional clients will grow as globalisation continues. At the same time, middle-sized firms, whether they are consultants or contractors, will continue to loose ground and fewer as well as larger organisations seek to provide one-stop services on an increasing scale. A growing number and range of relatively small specialists will have to be organised alongside in an effective manner, so as to best serve the needs of clients and their projects.

The tools presented here for a first selection of procurement routes and determining preferred business relationships between main contractors and their suppliers should be of help in increasingly dynamic and complex markets, without being either to prescriptive or complex as to prevent their use in every day practical situations.

It was discovered that producer-led procurement will not be appropriate under all circumstances, however, that the chances are very good for it to become increasingly more significant in all its forms of either Design and Build, Turn-key or BOT, as it offers clients significant benefits in tomorrows' markets.

References

Ambrose and Tucker, 2001.
Ambrose, M. D., and Tucker, S. N., Procurement System Evaluation for the Construction Industry; in: Journal of Construction Procurement, Vol. 6, No. 2, 2001, pp. 121-134.

Arthur Andersen and Enterprise LSE, 2001.
Artur Andersen and Enterprise LSE, Value for Money Drivers in the Private Finance Initiative, Report, commissioned by The Treasury Taskforce, 17th June 2001.

BAA, Dec. 2001.
British Airports Authority, Supplement, editor: Fairs, M.; in: Building, Issue 50, 14th Dec. 2001.

Barnes, 2001.
Barnes, C., Only yourself to blame; in: Building, Issue 35, 31st April 2001, p. 51.

Blecken, 1998.
Blecken, U., Die Kosten der öffentlichen Bauvorhaben; in: Bautechnik 75/3, 1998, pp. 180-187.

Booen, 2000.
Booen, P. L., The Three Major New FIDIC Books; in: The International Construction Law Review, Pt. 1, 2000, pp. 24-41.

Bremer and Kok, 2000.
Bremer, W. and Kok, K., The Dutch Construction Industry: a combination of competition and corporatism; in: Building Research & Information 28(2), 2000, pp. 98-108.

Brown, 2001.
Brown, D., Partnering: What happens when the team doesn't work?; in: Construction Information Quarterly (CIQ), Vol. 3, Issue 4, Construction Paper 137, pp. 13-21.

Brüssel, 1995.
Brüssel, W., Baubetrieb von A bis Z, 2nd ed., Düsseldorf, Werner Verlag, 1995, p. 1.

Building 11/1/2002.
Welcome to the Eurozone, Issue 1, 11th Jan. 2002, pp. 36-47.

Building 26/10/2001.
The specialists, Issue 43, 26th Oct. 2001, pp. 40-49.

Building 7/9/2001.
Amec: services switch pays off, Issue 36, 7th Sept. 2001, p. 23.

Building 8/2/2002.
Spanish firm agrees £ 462 m. for HBG, Issue 5, 8th Feb. 2002, p. 12.

Burchardt, 2001.
Burchardt, H. P., Die Arbeitsgemeinschaft (ARGE); in: Freiberger Handbuch zum Baurecht, editors: Jacob, D., Ring, G., Wolf, R., Bonn, Deutscher Anwalt Verlag and Berlin, Ernst & Sohn, 2001, pp. 857-876.

Campagnac, 2000.
Campagnac, E., The Contracting System in the French Construction Industry: acts and institutions; in: Building Research & Information 28(2), 2000, pp. 131-140.

Chevin, 1999.
Chevin, D., Spending power – The biggest-ever survey of what clients want from contractors; in: Building, Issue 47, 26th Nov. 1999, pp. 20-23.

CIOB, 1995.
Chartered Institute of Building, Financial Management of Building Contracts – 3: Sub-Contract Administration, Englemere, 1997.

CIOB, 1997 a).
Chartered Institute of Building, Code of Estimating Practice, 6th ed., Harlow, Addison Wesley Longmann, 1997.

CIOB, 1997 b).
Chartered Institute of Building, Financial Management of Building Contracts – 2: Selection of Sub-Contractors, Englemere, 1995.

CIOB, 1999.
Chartered Institute of Building, Code of Practice for Project Management for Construction, 2nd ed., Harlow, Pearson Education, 1999.

Construction Industry Board, 1997.
Construction Industry Board, Code of Practice for the Selection of Subcontractors, London, Thomas Telford, 1997.

Construction Industry Institute.
Construction Industry Institute, Project Delivery Systems: CM at risk, Design-Build, Design-Bid-Build, Research Survey, No. 133-1, 1997; in: Levey, S. M., Subcontractor's Operation Manual, New York, Mc-Graw Hill, 1999.

Cordis RTD, 2000.
Cordis-RTD, Acronyms, Record Control Nos. 1909, 5509, European Communities, Brussels, 2000.

Costantino, Pietroforte and Hamill, 2001.
Costantino, N., Pietroforte, R. and Hamill, P., Subcontracting in commercial and residential markets: an empirical investigation; in: Construction Management and Economics, Issue 19, 2001, pp. 439-447.

Cox and Townsend, 1998.
Cox, A. and Townsend, M., Strategic Procurement in Construction, London, Thomas Telford, 1998.

Davey, Lowe and Duff, 2001.
Davey, C. L., Lowe, D. J. and Duff, A. R., Generating opportunities for SMEs to develop partnerships and improve performance; in: Building Research & Information 29(1), 2001, pp. 1-11.

Davis Langdon & Everest, 2002 .
Davis Langdon and Everest, Cost model; in: Building, Issue 6, 15th Feb. 2002, pp. 67-72.

Devoy, 2001.
Devoy, F., When size is the prize; in: Building, Issue 12, 23rd March 2001, p. 47.

Dielschneider, 2000.
Dielschneider, J., Project Management in the US – Overview, Trends, Challenges; in: Latest Topics in Construction Business Management Winter 1999/2000, editors: Jacob, D., Winter C., Freiberg Working Papers, 28/2000, pp. 18-37.

Ernzen and Schexnayder, 2000.
Ernzen, J. J. and Schexnayder, G., One Company's Experience with Design/Build: Labour Cost Risk and Profit Potential; in: Journal of Construction Engineering and Management, Jan./Feb. 2000, pp. 10-14.

Fairs, 2001.
Fairs, M., Egan warns against turning to architects for advice; in: Building, Issue 49, 7th Dec. 2001, p. 18.

FIDIC, 1994.
Federation Internationale Des Ingenieurs-Conseils, Lausanne, 1994.

Franks, 1997.
Franks, J., Sub-Contract Conditions Associated with the JCT Intermediate Form of Building Contract – A view from the sub-contractor: 1, Construction Papers, No. 84, CIOB, 1997.
Franks, 1998.
Franks, J., Sub-Contract Conditions Associated with the JCT Intermediate Form of Building Contract – A view from the sub-contractor: 2, Construction Papers, No. 85, CIOB, 1998.

Gralla, 2001.
Gralla, M., Garantierter Maximalpreis: GMP-Partnering-Modelle – Ein neuer und innovativer Ansatz für die Baupraxis, Stuttgart, Teubner, 2001.

Gibb and Isack, 2001.
Gibb, A. G. F. and Isack, F., Client drivers for construction projects: implications for standardization; in:Engineering, Construction and Architectural Management, 8/1, Oxford, Blackwell Science, 2001, pp. 46-58.

Greenwood, 2001.
Greenwood, D., Subcontract procurement: are relationships changing?; in: Construction Management and Economics, No. 19, 2001, pp. 5-7.

Gruneberg and Ive, 2000.
Gruneberg, S. L. and Ive, G. J., The Economics of the Modern Construction Firm, London, MacMillan, 2000.

Halpin and Woodhead, 1998.
Halpin, D. W. and Woodhead, R. W., Construction Management, 2nd ed., New York, John Wiley, 1998.

Hamm, 1997.
Hamm, V., Informationstechnik-basierte Referenzprogramme, Univ. Diss., TU Freiberg, April 1997.

Hauptverband der deutschen Bauindustrie.
Hauptverband der deutschen Bauindustrie, Web-Elvira Datenbank, Zeitreihen: Arbeitsstudien, Umsatz, Beschäftigte und Betriebe, www.bauindustrie.de.

Helmus and Weber, 2000.
Helmus, M. and Weber, A., Wie hältst du's mit den Subs?; in: Bauwirtschaft, No. 5, 2000, pp. 30-33.

Herdt, 2000.
Herdt, C. M., Einsatz von Subunternehmern, Risiken und Vermeidung für mittelständische Bauunternehmen; in: Baumarkt, No. 7, 2000, pp. 21-25.

Hill, 2000.
Hill, R. M., Better Building: integrating the supply chain, a guide for clients and their consultants, Digest 450, Building Research Establishment (BRE), 2000.

Hofmann, 1992.
Hofmann, O., Die rechtliche Gestaltung von Subunternehmerverträge; in: Seminar: ARGE, GU, SU, Schriftenreihe der Deutschen Gesellschaft für Baurecht e. V., Vol. 19, Wiesbaden, Berlin, Bauverlag, 1992, pp. 66-75.

Horner, 1999.
Horner, R. M. W., Construction Project Management; Today's Challenges, Tomorrow's Opportunities, Construction Paper 99; in: Construction Information Quarterly, Vol.1, Issue 1, 1999, pp. 1-5.

Howell, 1999.
Howell, J., Where now for design & build?; in: Construction Manager, Vol. 5, Issue 5, CIOB, June 1999, pp. 34-36.

Hughes, Gray and Murdoch, 1997.
Hughes, W., Gray, C. and Murdoch, J., Specialist Trade Contracting – a Review, Specialist Publication 138, Construction Industry Research and Information Association (CIRIA), London, 1997.

Jacob and Kochendörfer, 2000.
Jacob, D. and Kochendörfer, B., Private Finanzierung öffentlicher Bauvorhaben – ein EU-Vergleich, Berlin, Ernst & Sohn, 2000.

Jacob, 1997.
Jacob, D., Aldi am Bau – Gedanken zu einer Analogie zwischen Handel und modernem Baugeschehen; in: Festschrift für Egon Heinrich Schlenke, Verband der Bauindustrie für Niedersachsen, Hannover, 1997, pp. 505-509.

Jacob, Winter and Stuhr, 2002.
Jacob, D., Winter, C. and Stuhr, C., Kalkulationsformen im Ingenieurbau, Berlin, Ernst & Sohn, 2002.

Kale and Arditi, 2001.
Kale S. and Arditi, D., General contractors' relationship with subcontractors: a strategic asset; in: Construction Management and Economics, Issue 19, 2001, pp. 541-549.

Kapellmann, 1997.
Kapellmann, K. D., Schlüsselfertiges Bauen – Rechtsgrundlagen zwischen Auftraggeber, Generalunternehmer, Nachunternehmer, 1st ed., Düsseldorf, Werner Verlag, 1997.

Klemmer, 1998.
Klemmer, J., Neustrukturierung bauwirtschaflicher Wertschöpfungsketten – Leistungstiefenoptimierung als strategisches Problemfeld, Wiesbaden, Deutscher Universitäts Verlag, 1998.

Kniffka, 1992.
Kniffka, R., Rechtliche Probleme des Generalunternehmervertrags; in: Seminar: ARGE, GU, SU, Schriftenreihe der Deutschen Gesellschaft für Baurecht e. V., Vol. 19, Wiesbaden, Berlin, Bauverlag, 1992, pp. 46-65.

Kochendörfer and Liebchen, 2001.
Kochendörfer, B. and Liebchen, J., Bau-Projekt-Management – Grundlagen und Vorgehensweisen, Stuttgart, B. G. Teubner, 2001.

Kommission der Europäischen Gemeinschaft, 1997.
Kommission der Europäischen Gemeinschaft, Die Wettbewerbsfähigkeit der Bauwirtschaft, Report, Brussels, Nov. 1997.

Kubal, Miller and Worth, 2000.
Kubal, M. T., Miller, K. T. and Worth, R. D., Building Profits in the Construction Industry, New York, Mc-Graw Hill, 2000.

Kumaraswamy and Matthews, 2000.
Kumaraswamy, M. M. and Matthews, J. D., Improved Subcontractor Selection Employing Partnering Principles; in: Journal of Management in Engineering, May/June 2000, pp. 47-57.

Lahdenperä, 2000.
Lahdenperä, P., Restructuring the Building Industry for Improved Performance; in: Journal of Construction Procurement, Vol. 5, No. 2, 2000.

Lamont, 2001 a).
Lamont, Z., Consider the client; in: Building, Issue 49, 7th Dec. 2001, p. 34.

Lamont, 2001 b).
Lamont, Z., Clients are loosing patience; in: Building, Issue 41, 12th Oct. 2001, p. 16.

Lampl, 2001.
Lampl, F., State your position, Issue 22, 1st June 2001, p. 47.

Latham, 1994.
Latham, M., Constructing the Team: Joint Review of Procurement and Contractual Arrangements in the United Kingdom Construction Industry, Final Report, London, The Stationary Office, July 1994.

Levey, 1999.
Levey, S. M., Subcontractor's Operation Manual, New York, Mc-Graw Hill, 1999.

Linden, 1999.
Linden, M., Risikomanagement gegen den Baustellenteufel; in: Bauwirtschaft; Issue 9, 1999, p. 9.

Ling, Khee and Lim, 2001.
Ling, Y. Y., Khee, H. Y. and Lim, K. S. G., The reason why clients prefer to procure more projects based on Design-Bid-Build than Design and Build; in: Journal of Construction Procurement, Vol. 6, No. 2, 2001, pp. 135-145.

Ling, Ofori and Lam, 2000.
Ling, Y. Y., Ofori, G. and Lam, S. P., Importance of design consultants' soft skills in design-based Projects; in: Engineering, Construction and Architectural Management, 7/4, Oxford, Blackwell Science, 2000, pp. 389-398.

Loh and Ofori, 2000.
Loh, W. H. and Ofori, G., Effect of registration on performance of construction subcontractors in Singapore; in: Engineering, Construction and Architectural Management, 7/1, London, Blackwell Science, 2000, pp. 29-40.

Madine, 2001 a).
Madine, V., Engineering change; in: Building, Issue 26, 29th June 2001, pp. 42-43.

Madine, 2001 b).
Madine, V., Boots looks for firms to run new prime contract regime; in: Building, Issue 27, 6th July 2001, p. 17.

Male, et al., 1998 a).
Male, S., et al., The Value Management Benchmark: A good practice framework for clients and practitioners, London, Thomas Telford, 1998.

Male, et al., 1998 b).
Male, S., et al., The Value Management Benchmark: Research results of an international benchmarking study, London, Thomas Telford, 1998.

Masters, 1998.
Masters, J., Taxing problems overcome; in: Construction Manager, Vol. 4, Issue 5, CIOB, June 1998, pp. 10.

McAll, 2000.
McAll, D., Merger.com; in: Construction Computing, Dec. 2000, pp. 12-13.

McCaffer and Baldwin, 1995.
McCaffer, R. and Baldwin, A., Estimating for construction; in: Project Cost Estimating, editor: Smith, N. J., London, Thomas Telford, 1995, pp. 34-50.

Medicus, 1992.
Medicus, D., Abnahme und Gewährleistung im Verhältnis Generalunternehmer – Subunternehmer; in: Seminar: ARGE, GU, SU, Schriftenreihe der Deutschen Gesellschaft für Baurecht e. V., Vol. 19, Wiesbaden, Berlin, Bauverlag, 1992, pp. 76-85.

Mehrtens, 1996.
Mehrtens, H. A. J., Abwicklung von Vertragsleistungen in Arbeitsgemeinschaften mit Nachunternehmern oder Fremdarbeitern aus Sicht einer mittelständischen Firma, Presentation, Köln, 12th June 1996.

Merna and Smith, 1996 a).
Merna, A. and Smith, N. J., Projects procured by Privately Financed Concession Contracts, Vol. 1, 2nd ed., Hong Kong, Asia Law and Practice, 1996.

Merna and Smith, 1996 b).
Merna, A. and Smith, N. J., Projects procured by Privately Financed Concession Contracts, Vol. 2, Hong Kong, Asia Law and Practice, 1996.

National Audit Office, 2001.
National Audit Office, Managing the relationship to secure a successful partnership in PFI projects, Report, HC 375, London, Stationary Office, 29th Nov. 2001, www.nao.gov.uk.

Newcombe, 2001.
Newcombe, R., An Investigation into simulating the Procurement Process in the UK Construction Industry; in: Journal of Construction Procurement, Vol. 6, No. 2, 2001, pp. 104-120.

Office of Government Commerce, 2000.
Office of Government Commerce (OGC), Value for Money Measurement; OGC Business Guide, London, Nov. 2000.

Passarge and Warner, 2001
Passarge, J. and Warner, M, Privates Baurecht; in: Freiberger Handbuch zum Baurecht, editors: Jacob, D., Ring, G., Wolf, R., Bonn, Deutscher Anwalt Verlag and Berlin, Ernst & Sohn, 2001, pp. 11-372.

Pettinger, 1998.
Pettinger, R., Construction Marketing – Strategies for Success, London, Macmillan, 1998.

Picot, 1991.
Picot, A., Ein neuer Ansatz zur Gestaltung der Leistungstiefe; in: Zeitschrift für betriebswirtschaftliche Forschung, No. 43, 1991, pp. 336-357.

Pilcher, 1997.
Pilcher, R., Principles of Construction Management, 3rd ed., London, Mc-Graw Hill, 1997.

Pisoni, 2001.
Pisoni, M., Möglichkeiten der Beschaffung von Baustoffen und Nachunternehmer-leistungen im italienschen Wirtschaftsraum, Univ. Dipl., TU Freiberg, July 2001.

Porter, 1985.
Porter, M. E., Competitive Advantage, creating and sustaining superior performance, New York, Free Press, 1985.

Porter, 1990.
Porter, M. E., The Competitive Advantage of Nations, London, MacMillan, 1990.

Puddicombe, 1997.
Puddicombe, M. S., Designers and Contractors: Impediments to Integration; in: Journal of Construction Engineering and Management, Vol. 123, No. 3, Sept. 1997, pp. 245-252.

Ramsey, 1996 a).
Ramsey, J., The Case Against Purchasing Partnerships; in: International Journal of Purchasing and Materials Management, Nov. 1996, pp. 13-19.

Ramsey, 1996 b).
Ramsey, J., Partnerships of unequals; in: Supply Management, 28th March 1996, pp. 31-33.

Rösel, 1994.
Rösel, W., Baumanagement – Grundlagen, Technik, Praxis, 3rd ed., Berlin, Springer, 1994.

Ruckteschler, 1988.
Ruckteschler, D., Subunternehmerhaftung: Möglichkeiten des Durchgriffs von Bauherrn und Käufern nach amerikanischen, englischen und deutschen Recht, Europäische Handelsschriften, 2/755, Frankfurt/Main, 1988.

Sash, 1998.
Sash, A. A., Bidding Practices of Subcontractors in Colorado; in: Journal of Construction Engineering and Management, Vol. 124, No. 3, ASCE, May/June 1998, pp. 219-225.

Schulze, 1997.
Schulze, D., Informationstechnik-basierte Referenzprogramme konkret bezogen auf den Einkaufsprozeß eines Bauunternehmens, Univ. Dipl., TU Berlin, Oct. 1997.

Schwarz and Schmutzer, 1997.
Schwarz, S. and Schmutzer, M. O., Zentrale Marktentwicklungen in der Bauwirtschaft; in: Zukunftssicherung für die Bauwirtschaft – in vier Schritten aus der Krise, Wiesbaden, Gabler, 1997, pp. 13-21.

Schwarz, 1996.
Schwarz, S., Der Subunternehmervertrag, Univ. Diss., Hamburg, 1996.

Seddon, 2001.
Seddon, E., Meet our man in London, Paris, New York ...; in: Building, Issue 12, 23rd March 2001, pp. 44-46.

Seely, 1997.
Seely, I. H., Quantity Surveying Practice, 2nd ed., London, MacMillan, 1997.

Simm, 2000.
Simm, J., Bid Procurement: one contractor's view; in: CES Journal, June 2000, pp. 28-30.

Smith, 1995.
Smith, N. J., Engineering Project Management,Oxford, Blackwell Science, 1997.

Sommer, 2000.
Sommer, H., Projektmanagement im Hochbau, 2nd ed., Berlin, Springer, 2000.

Sözen and Kayahan, 2001.
Sözen, Z. and Kayahan, O., Correlates of the lenght of the relationships between main and specialist trade contractors in the construction industry; in: Construction Management and Economics, No. 19, 2001, pp. 131-133.

Sperling, 1999.
Sperling, W., Tendenzen in der Entwicklung der Bauunternehmen und ihr Zusammenhang zum Leistungsumfang, zum Preistyp und zur Erfolgssicherung; in: Ehrenkolloqium zum 65. Geburtstag von Herrn Prof. Dr. Ing. habil. J. Schindler, TU Dresden, 1999, pp. 25-35.

Stirling, 2001.
Stirling, J., Clobbered from the start; in: Building, Issue 26, 29th June 2001, p. 50.

Stumpf, 2000.
Stumpf, I., Competitive pressures on middle-market contractors in the UK; in: Engineering, Construction and Architectural Management, 7/2, Oxford, Blackwell Science, 2000, pp. 159-168.

Syben, 2000.
Syben, G., Contractors take command: from a demand based towards a producer orientated model in German construction; in: Building Research & Information, 28(2), 2000, pp. 109-130.

The Building Centre Trust, 2001.
The Building Centre Trust, Effective integration of IT in Construction, Report, London, 2001, www.buildingcentretrust.org.

The KPI Working Group, 2000.
The KPI Working Group, Report for the Minister for Construction, Department of the Environment, Transport and the Regions, London, June 2000, www.detr.gov.uk.

Thompson, 2001.
Thompson, G. J., Oscar Faber bought by US engineering giant Aecom; in: Building, Issue 42, 19th Oct. 2001, p. 23.

Thompson, 2002.
Thompson, G. J., Cream of the continent; in: Building, Issue 1, 11th Jan. 2002, pp. 50-61.

Tookey, Murray, Hardcastle and Langford, 2001.
Tookey, J. E., Murray, M., Hardcastle, C. and Langford, D., Construction Procurement Routes: re-defining the contours of construction procurement; in: Engineering, Construction and Architectural Management, 8/1, Oxford, Blackwell Science, 2001, pp. 20-30.

Turner, 2000.
Turner, J., The Evolution of the European Construction Market – An Economist's Perspective; in: Go Europe Go! Offensive für das Bauen in Europa – Kooperationen und strategische Allianzen, Tagungsband, bautec 16th-19th Feb., Berlin, 2000.

Walker, 1996.
Walker A., Project Management in Construction, 3rd ed., London, Blackwell Science, 1996.

Walter, 1998.
Walter, M., The Essential Accessory; in: Construction Manager, CIOB, Vol. 4, Issue 1, Feb. 1998, pp. 16-17.

Watson and Speak, 2001.
Watson P. and Speak, D., An Update on Construction Management Procurement; in: Building, Engineer, Feb. 2001, pp. 23-26.

Winter and Preece, 2000.
Winter, C. and Preece, C., Relationship Marketing between Specialist Subcontractors and Main Contractors – comparing UK and German practice; in: International Journal for Construction Marketing, Vol. 2, Issue 1, Oxford, Oxford Brookes University, Nov. 2000.

Winter, 2000.
Winter, C., Das Berufsbild des englischen Qunatity Surveyors – Derzeitige englische und zukünftige deutsche baubetriebswirtschaftliche Ausbildung; in: Latest Topics in Construction Business Management Summer 1999, editor: Jacob, D., Freiberg Working Papers, 2/2000, pp. 63-74.

Wischhof, 2000.
Wischhof, et al., Strategie für mittelständische Bauunternehmen in Europa, Rationalisierungs - Gemeinschaft der Deutschen Wirtschaft e. V. (RKW), Eschborn, RKW Verlag, 2000.

Wlasak, 2001.
Wlasak, P., Probleme des Wirtschaflichkeitsvergleich bei privatwirtschaftlich durchgeführten Bauinvestitionen; in: 2. Europäisches Symposium - Effizienzvorteile bei der privatwirtschaftlichen Realisierung öffentlicher Bauvorhaben, Berlin, Sept. 2001, pp. 59-65.

Wong and Fung, 1999.
Wong, A. and Fung, P., Total quality management in the construction industry in Hong Kong: a supply chain management perspective; in: Total Quality Management, Vol. 10, No. 2, 1999, pp. 199-208.

Wong, Holt and Cooper, 2000.
Wong, C. H., Holt, G. D. and Cooper, P. A., Lowest Price or Value? Investigation of UK construction clients' tender selection process; in: Construction Management and Economics, Issue 18, 2000, pp. 767-774.

Appendix

type	criteria	priority	value	designer-led single-stage	two-stage	serial tenders	separate trades	management-led CM for fee	CM at risk	MC for fee	MC at risk	producer-led Dev. & Const. D & B	Competitive D & B	Direct D & B	Turnkey/Package-Deal	BOT
A	price certainty	Do you need to have a firm price for as much of the procurement process as possible before you can commit	yes											■	■	■
			budget only							■						
B	timing	How important is early completion to the success of your project?	crucial											■	■	■
			important													
			not important													
C	controllable variation	Do you foresee the need to alter the project in any way once it has started on site?	yes													
			some													
			no													
D	complexity	Is your building of a high design or technical standard and can the project environment be described as dynamic, moderately so or not dynamic?	yes													
			moderately													
			no													
E	quality level	What level of quality (standard) do you seek in the design and workmanship of your project?	basic													
			good													
			prestige													
F	contractor input	How important is the ability to involve contractors' expertise at the design stage?	important													
			not important													
G	competition	Do you need to choose your construction team and/or work contractors by price competition?	work contractors													
			wks. & const. mgt. teams													
H	management	Can you manage many separate consultants and contractors, some, or do you want just one firm to be responsible after the briefing stage?	many separate forms													
			some separate firms													
			one firm only													
			no													
I	accountability	Do you want direct professional accountability to you from the designers and cost consultants?	no													
			yes													
J	risk avoidance	Do you want to pay someone to take the risk of cost and time slippage for you?	no													
			share													
			yes													
K	operation & maintenance	Do you want to pay someone to take the responsibility not only for designing and building, but for the operation and maintenance of your building as well?	no													
			share													
			yes													
		TOTALS		3	4	4	3	2	6	3	7	6	8	10	11	11

No. 1: Industrial producer (factory)

type	criteria	priority	answer options
A	price certainty	Do you need to have a firm price for as much of the procurement process as possible before you can commit	yes / budget only
B	timing	How important is early completion to the success of your project?	crucial / important / not important
C	controllable variation	Do you foresee the need to alter the project in any way once it has started on site?	yes / some / no
D	complexity	Is your building of a high design or technical standard and can the project environment be described as dynamic, moderately so or not dynamic?	yes / moderately / no
E	quality level	What level of quality (standard) do you seek in the design and workmanship of your project?	basic / good / prestige
F	contractor input	How important is the ability to involve contractors' expertise at the design stage?	important / not important
G	competition	Do you need to choose your construction team and/or work contractors by price competition?	work contractors / wks. & const. mgt. teams
H	management	Can you manage many separate consultants and contractors, some, or do you want just one firm to be responsible after the briefing stage?	many separate forms / some separate firms / one firm only
I	accountability	Do you want direct professional accountability to you from the designers and cost consultants?	no
J	risk avoidance	Do you want to pay someone to take the risk of cost and time slippage for you?	yes / no / share
K	operation & maintenance	Do you want to pay someone to take the responsibility not only for designing and building, but for the operation and maintenance of your building as well?	yes / no / share / yes

Procurement methods evaluated (with totals):

group	method	total
producer-led	BOT	6
producer-led	Turnkey/Package-Deal	6
producer-led	Direct D & B	7
producer-led	Competitive D & B	6
producer-led	Dev. & Const. D & B	9
management-led	MC at risk	9
management-led	MC for fee	9
management-led	CM at risk	8
management-led	CM for fee	6
designer-led	separate trades	6
designer-led	serial tenders	7
designer-led	two-stage	9
designer-led	single-stage	8

No. 2: Up-market food retailer (supermarket)

The table below is rotated on the page. Column headers (left to right across the data columns) are the value-labels; the procurement methods are the rows.

type	criteria	priority	yes	budget only	crucial	important	not important	yes	some	no	yes	moderately	no	basic	good	prestige	important	not important	work contractors	wks. & const. mgt. teams	many separate forms	some separate firms	one firm only	no	yes	no	share	yes	no	share	yes	TOTALS
producer-led BOT			•		•	•			•				•			•							•		•		•			•	2
Turnkey/Package-Deal			•		•	•			•				■		•	■							•		•	•				■	2
Direct D & B			•		•	•			•	•			■		•	■		•		•			•		■			•	■		3
Competitive D & B			•			•			•				■		•	•		•		•			•		■		•		■		3
Dev. & Const. D & B			•			•		•			•	•		•		■		■		•			•		•		■		■		5
management-led MC at risk			•		•	•		•	•	■		•			•		•	■		•	•			•			■		■		5
MC for fee		■			•	■		•	■		•			•			■				•		■			•	■				7
CM at risk			•		•	•		•	■		•				•		■			•	■			•			■		■		5
CM for fee			•		•	•		■	•		•				•		•			■	■			•			■		■		8
designer-led separate trades				■	■	•		•	■	•		■		•	•		•	■		■			■	■			■		■		11
serial tenders				■	•				■	•	•	■		•			•	■	•				•		•		■		■		6
two-stage				■	•	•		•	■	•	•	■		•	■			•		•			■		•		■		■		6
single-stage		■		■			■	•	■	•	•	■		•	■		•	■	•	•			■		•		■		■		8

type	criteria	priority
A	price certainty	Do you need to have a firm price for as much of the procurement process as possible before you can commit
B	timing	How important is early completion to the success of your project ?
C	controllable variation	Do you foresee the need to alter the project in any way once it has started on site ?
D	complexity	Is your building of a high design or technical standard and can the project environment be described as dynamic, moderately so or not dynamic ?
E	quality level	What level of quality (standard) do you seek in the design and workmanship of your project ?
F	contractor input	How important is the ability to involve contractors' expertise at the design stage ?
G	competition	Do you need to choose your construction team and/or work contractors by price competition ?
H	management	Can you manage many separate consultants and contractors, some, or do you want just one firm to be responsible after the briefing stage ?
I	accountability	Do you want direct professional accountability to you from the designers and cost consultants ?
J	risk avoidance	Do you want to pay someone to take the risk of cost and time slippage for you ?
K	operation & maintenance	Do you want to pay someone to take the responsibility not only for designing and building, but for the operation and maintenance of your building as well ?

No. 3.1: Local authority „traditional" (administration office)

	A: yes	A: budget only	B: crucial	B: important	B: not important	C: yes	C: some	C: no	D: yes	D: moderately	D: no	E: basic	E: good	E: prestige	F: important	F: not important	G: work contractors	G: wks. & const. met. teams	H: many separate forms	H: some separate firms	H: one firm only	I: no	I: yes	J: no	J: share	J: yes	K: no	K: share	K: yes	TOTALS
producer-led — BOT	■		•		■			■			■	■	•	•	■		■			■	•		■	■			■	•	■	11
Turnkey/Package-Deal	■		•		■			■			■	■	•	•	■		■			■	•		■	■			■	•	•	10
Direct D & B	■		•		■		•			•	■	■	•	•	■		■			•			■	■	•					9
Competitive D & B	■				■			■			■	■	•	•	■		■			■			■	■	•					10
Dev. & Const. D & B	■		•		•			■		■	•	•	•	■	•		■			•			•	■	•					8
management-led — MC at risk	■		•		■		■	•	■		•	■	•	•	■		■			•	•		•	■	•					8
MC for fee		•			■		•	•		•	■	■	•	•	■		•			•	•		•	•	•					5
CM at risk	■		•		•		•	•		•	■	■	•	•	•		■			•	•		•	■	•					7
CM for fee		•			■		•	•		•	■		•	•	■		•			•	•		•	•						4
designer-led — separate trades		•			•		•	•		■	•	•			■	•				•	•		•	•						4
serial tenders		•		■				■		■	•	•			•	•				•	•		•	•						4
two-stage		•			■		•	•		■	•	■			■	•				•	•		•	•	•					6
single-stage		•			•		•	•		■	•	■			•	•				•	•		•	•						4

type	criteria	priority
A	price certainty	Do you need to have a firm price for as much of the procurement process as possible before you can commit
B	timing	How important is early completion to the success of your project?
C	controllable variation	Do you foresee the need to alter the project in any way once it has started on site?
D	complexity	Is your building of a high design or technical standard and can the project environment be described as dynamic, moderately so or not dynamic?
E	quality level	What level of quality (standard) do you seek in the design and workmanship of your project?
F	contractor input	How important is the ability to involve contractors' expertise at the design stage?
G	competition	Do you need to choose your construction team and/or work contractors by price competition?
H	management	Can you manage many separate consultants and contractors, some, or do you want just one firm to be responsible after the briefing stage?
I	accountability	Do you want direct professional accountability to you from the designers and cost consultants?
J	risk avoidance	Do you want to pay someone to take the risk of cost and time slippage for you?
K	operation & maintenance	Do you want to pay someone to take the responsibility not only for designing and building, but for the operation and maintenance of your building as well?

No. 3.2: Local authority „transfer of risk to private party" (administration office)

type	criteria	priority	yes	budget only	crucial	important	not important	yes	some	no	yes	moderately	no	basic	good	prestige	important	not important	work contractors	wks. & const. mgt. teams	no	many separate forms	some separate firms	one firm only	no	yes	no	share	yes	no	share	yes	TOTAL
		producer-led																															
		BOT	•			•	■				•					•	■				•			•		•			•		•	•	2
		Turnkey/Package-Deal	•			•	■				•					•	■				•			•		•		•	•		•	•	2
		Direct D & B	•			•	■					•	•				■				•			•		•		•	•	■		•	3
		Competitive D & B	•			■				•					•	•	■		•				•	•		•		•	•	■		•	3
		Dev. & Const. D & B	•			■			•	•		•	•		•		■		■				■			■		•	■		•		6
		management-led																															
		MC at risk	•			•	■		•	•		•	■		•	•	■		■	•			■			■		•	■		•	•	6
		MC for fee		■		■		■			■	•		■	•		■		■	•			■		■		■		•	■		•	10
		CM at risk		•		•	■		•	•		■			•		■	■	•	■		■			•		■	•		■		•	7
		CM for fee		•		•	■		•	•		■		■	•		■	■	•	■			•		■		•	■		•		•	10
		designer-led																															
		separate trades		■		■		•	■	•		•			•		■		•			■			■		•	■		■		•	6
		serial tenders		■		■			•			•			•		■	•				•			■		■	■		■		•	5
		two-stage		■		■			•			•			•		■	•	•			•			■		■	■		■		•	7
		single-stage	■			•		•	•		•	•		•	•		•	•	•			■			■		■	■		■		•	5
		TOTALS	5	7	5	7	6	10	7	10	6	3	3	3	2	2																	

type / criteria / priority

- **A — price certainty:** Do you need to have a firm price for as much of the procurement process as possible before you can commit
- **B — timing:** How important is early completion to the success of your project?
- **C — controllable variation:** Do you foresee the need to alter the project in any way once it has started on site?
- **D — complexity:** Is your building of a high design or technical standard and can the project environment be described as dynamic, moderately so or not dynamic?
- **E — quality level:** What level of quality (standard) do you seek in the design and workmanship of your project?
- **F — contractor input:** How important is the ability to involve contractors' expertise at the design stage?
- **G — competition:** Do you need to choose your construction team and/or work contractors by price competition?
- **H — management:** Can you manage many separate consultants and contractors, some, or do you want just one firm to be responsible after the briefing stage?
- **I — accountability:** Do you want direct professional accountability to you from the designers and cost consultants?
- **J — risk avoidance:** Do you want to pay someone to take the risk of cost and time slippage for you?
- **K — operation & maintenance:** Do you want to pay someone to take the responsibility not only for designing and building, but for the operation and maintenance of your building as well?

No. 4. Financial institution (new headquarters)

Criteria, priority questions and answer options

type	criteria	priority	answer options
A	price certainty	Do you need to have a firm price for as much of the procurement process as possible before you can commit	yes / budget only
B	timing	How important is early completion to the success of your project?	crucial / important / not important
C	controllable variation	Do you foresee the need to alter the project in any way once it has started on site?	yes / some / no
D	complexity	Is your building of a high design or technical standard and can the project environment be described as dynamic, moderately so or not dynamic?	yes / moderately / no
E	quality level	What level of quality (standard) do you seek in the design and workmanship of your project?	basic / good / prestige
F	contractor input	How important is the ability to involve contractors' expertise at the design stage?	important / not important
G	competition	Do you need to choose your construction team and/or work contractors by price competition?	work contractors / wks. & const. mgt. teams
H	management	Can you manage many separate consultants and contractors, some, or do you want just one firm to be responsible after the briefing stage?	many separate firms / some separate firms / one firm only
I	accountability	Do you want direct professional accountability to you from the designers and cost consultants?	no / yes
J	risk avoidance	Do you want to pay someone to take the risk of cost and time slippage for you?	no / share / yes
K	operation & maintenance	Do you want to pay someone to take the responsibility not only for designing and building, but for the operation and maintenance of your building as well?	no / share / yes

Procurement methods and TOTALS

group	method	TOTAL
producer-led	BOT	10
producer-led	Turnkey/Package-Deal	9
producer-led	Direct D & B	8
producer-led	Competitive D & B	8
producer-led	Dev. & Const. D & B	7
management-led	MC at risk	8
management-led	MC for fee	4
management-led	CM at risk	6
management-led	CM for fee	2
designer-led	separate trades	4
designer-led	serial tenders	4
designer-led	two-stage	6
designer-led	single-stage	5

No. 5: Developer (basic office space)

Deutscher Universitäts-Verlag

Ihr Weg in die Wissenschaft

Der Deutsche Universitäts-Verlag ist ein Unternehmen der Fachverlagsgruppe BertelsmannSpringer, zu der auch der Gabler Verlag und der
Vieweg Verlag gehören. Wir publizieren ein umfangreiches wirtschafts-
wissenschaftliches Monografien-Programm aus den Fachgebieten

✓ Betriebswirtschaftslehre
✓ Volkswirtschaftslehre
✓ Wirtschaftsrecht
✓ Wirtschaftspädagogik und
✓ Wirtschaftsinformatik

In enger Kooperation mit unseren Schwesterverlagen wird das Pro-
gramm kontinuierlich ausgebaut und um aktuelle Forschungsarbeiten
erweitert. Dabei wollen wir vor allem jüngeren Wissenschaftlern ein
Forum bieten, ihre Forschungsergebnisse der interessierten Fachöffent-
lichkeit vorzustellen. Unser Verlagsprogramm steht solchen Arbeiten
offen, deren Qualität durch eine sehr gute Note ausgewiesen ist. Jedes
Manuskript wird vom Verlag zusätzlich auf seine Vermarktungschancen
hin geprüft.

Durch die umfassenden Vertriebs- und Marketingaktivitäten einer gro-
ßen Verlagsgruppe erreichen wir die breite Information aller
Fachinstitute, -bibliotheken und -zeitschriften. Den Autoren bieten wir
dabei attraktive Konditionen, die jeweils individuell vertraglich verein-
bart werden.

Besuchen Sie unsere Homepage: *www.duv.de*

Deutscher Universitäts-Verlag
Abraham-Lincoln-Str. 46
D-65189 Wiesbaden